Lecture Notes in Mathematics

Edited by A. Dold and B. Eckmann

451

Probabilistic Methods in Differential Equations

Proceedings of the Conference Held at the University of Victoria, August 19–20, 1974

Edited by M. A. Pinsky

Springer-Verlag
Berlin · Heidelberg · New York 1975

Editor
Dr. Mark A. Pinsky
Department of Mathematics
Northwestern University
College of Arts and Sciences
Evanston, Illinois 60201
USA

Library of Congress Cataloging in Publication Data

Conference on Probabilistic Methods in Differential
Equations, University of Victoria, 1974.
Probabilistic methods in differential equations.

(Lecture notes in mathematics ; 451)
Bibliography: p.
Includes index.
1. Stochastic differential equations--Congresses.
2. Markov processes--Congresses. I. Pinsky, Mark A., 1940- II. Title. III. Series: Lecture notes in mathematics (Berlin) ; 451)
QA3.L28 no. 451 [QA274.23] 510'.8s [519.23] 75-12982

AMS Subject Classifications (1970): 47 D 05, 60 F 05, 60 F 10, 60 G 05, 60 G 40, 60 H 10, 60 H 15, 60 J 25, 60 J 35, 60 J 55, 60 J 60, 60 J 65, 60 J 80

ISBN 3-540-07153-9 Springer-Verlag Berlin · Heidelberg · New York
ISBN 0-387-07153-9 Springer-Verlag New York · Heidelberg · Berlin

Printed in Germany

Offsetdruck: Julius Beltz, Hemsbach/Bergstr.

INTRODUCTION

The Conference on Probabilistic Methods in Differential Equations was held in Victoria, British Columbia, August 19-20, 1974. The purpose of the conference was to bring together workers in the field of stochastic differential equations and closely related areas. With a total of 35-40 participants, it was possible to maintain an air of informality and active mathematical interchange during the two-day conference.

There were 4 one-hour lectures and 10 twenty-minute talks presented at the conference. As several of the talks consisted of reports on work published elsewhere, only 11 complete manuscripts appear here. The remaining 3 talks are listed by title. The talks are listed in the order in which they were presented at the conference.

We gratefully acknowledge the financial support of the National Research Council of Canada, Grant # 1550-98. The local arrangements were made by C. Robert Miers of the Mathematics Department, University of Victoria, without whose tireless efforts the conference would not have been possible. Finally, we wish to acknowledge a travel grant from the Northwestern University Research Committee.

M. Pinsky

Evanston, Illinois

December 31, 1974

CONFERENCE ON

PROBABILISTIC METHODS IN DIFFERENTIAL EQUATIONS

List of Participants

AIRAULT, Helene
Université Paris
Paris, France.

BAJAJ, Prem N.
Dept. of Math.,
Wichita State University
Wichita, Kansas 67208

BENSOUSSAN, Alain,
IRIA Domaine de Voluceau
Université Paris IX
Le Chesnay, France

BURKHOLDER, D. L.
Dept of Math.
University of Illinois
Urbana, Illinois 61801

CAUBET, Jean-Pierre
Dept. of Math.
Université de Poitiers
Poitiers, France

COOKE, Kenneth L.
Dept. of Math.
Pomona College
Claremont, California 91711

COPPEL, W. A.
Dept. of Math.
Australian National University
Canberra, Australia.

DAWSON, Donald A.
Carleton University
Ottawa, Ontario, K1S 5B6

DERRICK, William R.
Dept. of Math.
University of Montana,
Missoula, Montana 59801

ELLIS, Richard S.
Dept. of Math
Northwestern University
Evanston, Illinois 60201

ELWORTHY, K. D.
Maths Institute,
University of Warwick
Coventry, England

FRIEDMAN, Avner
Dept Math.
Northwestern University
Evanston, Illinois 60201

GETOOR, R. K.
Dept. of Math
University of California,
San Diego
La Jolla, California 92037

BEN-GHANDOUR, Addi
Dept. of Math. Sciences
University of Tel-Aviv
Ramat-Aviv, Israel.

GODDARD, L. S.
Dept. of Math.
University of Salford
Lancs., U.K.

GOODMAN, Victor
Dept. of Math.
Indiana University
Bloomington, Indiana 47401

GREENWOOD, Priscilla
Dept. of Math.
University of British Columbia
Vancouver, B. C.

GRIEGO, Richard J.
Dept. of Math.
University of New Mexico
Albuquerque, New Mexico 87106

HAHN, Marjorie G.
M.I.T.
Cambridge, Mass. 02142

HERSH, R.
Math. Dept.
University of New Mexico
Albuquerque, New Mexico 87106

INDELLI, Paola
M.I.T.
Cambridge, Mass. 02142

ITÔ, Kiyosi
Dept. of Math.
Cornell University
Ithaca, New York 14850

JOSHI, C. M.
Dept. of Math.
University of Jodhpur
Jodhpur, India

KURSS, Herbert
Dept. of Math
Adelphi, University
Garden City, L.I. N.Y.11530

KURTZ, Thomas G.
Dept. of Math
University of Wisconsin
Madison, Wisconsin53706

LARA-CARRERO, Lorenzo
M.I.T. & IVIC (Venezuela)
Cambridge, Mass.

MCKELVEY, Robert
Dept. of Math.
University of Montana
Missoula, Montana 59801

MALIK, M.A.
Dept. of Math.
Sir George Williams University
Montreal

MALLIAVIN, Paul
School of Mathematics
The Institute for Advanced Study
Princeton, New Jersey

MARKUS, Lawrence
School of Mathematics
University of Minnesota
Minneapolis, Minn.

MONROE, Itrel
Dept. of Math
University of Arkansas
Fayetteville, Arkansas

PAPANICOLAOU, George C.
Courant Institute
New York, N. Y. 10012

PINSKY, Mark A.
Dept. of Math.
Northwestern University
Evanston, Illinois 60201

ROSENBLATT, Murray
Dept. of Math.
University of California, San Diego
La Jolla, California

SAWYER, Stanley
Belfer Graduate School
Yeshiva University
New York, N.Y. 10033

TAYLOR, W. Clare
3005 Snake Lane
Churchville, Maryland

VARADHAN, S.R.S.
Courant Institute
New York University
New York, N.Y. 10012

WANG, Frank J. S.
Dept. Math.
University of Montana
Missoula, Montana 59801

WELLAND, Grant V.
Dept. of Math.
University of Missouri
St. Louis, Missouri 63121

CONTENTS

* These were one-hour lectures.

† These speakers did not submit manuscripts.

Stochastic parallel displacement

Kiyosi Itô*

1. Introduction. In our previous paper [1] we have introduced a stochastic differential dX as a random interval function induced from a continuous local quasi-martingale X. If $F(x_1,x_2,\ldots,x_n)$ is sufficiently smooth, we have a chain rule

$$(C) \qquad dF(X_1,X_2,\ldots,X_n) = \sum_i \partial_i F\, dX_i + \frac{1}{2}\sum_{i,j} \partial_i\partial_j F\, dX_i\, dY_j$$

where

$$(A\,dX)(I) = \int_I A\,dX \qquad \text{(stochastic integral)}$$

and $dX\,dY = d(XY) - XdY - YdX$ (quadratic covariation).
If we use the symmetric multiplication

$$A\circ dX = AdX + \frac{1}{2}\,dAdX$$

which corresponds to the Fisk-Stratonovich symmetric stochastic integral [5][6], the chain rule (C) can be written as

$$(C_s) \qquad dF(X_1,X_2,\ldots,X_n) = \sum_i \partial_i\, F\circ dX_i .$$

Since (C_s) takes the same form as in the ordinary calculus, the symmetric multiplication is convenient for some purpose. We have given such examples in our previous paper [1]. In the present paper we will discuss stochastic parallel displacement as another interesting example.

*Supported by NSF GP-33136X, Cornell University.

2. Stochastic parallel displacement. Let us review some notation in differential geometry following Dynkin [3]. Let $S = (S, \Gamma^i_{jk})$ be an affinely connected ℓ-dimensional C^3-manifold and $T^n_m = (T^n_m(x), x \in S)$ the bundle of tensors of type (m,n). $T^n_m(x)$ is dual to $T^m_n(x)$ relative to the invariant bilinear form:

(1) $$(u,w) := u^{j_1\cdots j_n}_{i_1\cdots i_m} w^{i_1\cdots i_m}_{j_1\cdots j_n} ,$$

where the summation sign is omitted as is common in differential geometry. Using the symbol

(2) $$(\Gamma_i u)^{j_1\cdots j_n}_{i_1\cdots i_m} = \Gamma^k_{ii_\mu} u^{j_1\cdots j_n}_{i_1\cdots i_{\mu-1}k\, i_{\mu+1}\cdots i_m} - \Gamma^{j_\nu}_{ik} u^{j_1\cdots j_{\nu-1}k j_{\nu+1}\cdots j_n}_{i_1\cdots i_m} ,$$

we can express the covariant derivative ∇_i as follows:

(3) $\nabla_i = \partial_i - \Gamma_i$, where $\partial_i = \dfrac{\partial}{\partial x_i}$.

Let $C: y(t)$, $t_0 \leq t \leq t_1$, be a smooth curve on S. The tensor $u_1 \in T^n_m(y(t_1))$ is called parallel to $u_0 \in T^n_m(y(t_0))$ along C, if u_0 and u_1 are connected by a family of tensors $u(t) \in T^n_m(y(t))$, $t_0 \leq t \leq t_1$ satisfying

(4) $\dot{u} = (\Gamma_i u)\dot{y}^i$, i.e. $du = (\Gamma_i u)dy^i$, $i = 1,2,\ldots,\ell$.

Now we want to define the stochastic parallelism along a random curve $C(\omega)$: $Y_t(\omega)$, $t_0 \leq t \leq t_1$. Since the sample curve of $C(\omega)$ is not smooth in general, we cannot apply the above definition to each sample curve, but we can define stochastic parallelism by replacing the equation (4) with its stochastic analogue:

$$(5) \qquad dU = (\Gamma_i U)\circ dY^i,$$

where the small circle denotes the symmetric multiplication of stochastic differentials [1], so (5) is also expressed as

$$(5') \qquad dU = (\Gamma_i U)dY^i + 1/2d(\Gamma_i U)dY^i.$$

We can equivalently define stochastic parallelism in the following geometric way. Let $U_1(\omega)$ be a random tensor in $T^n_m(Y(t_0,\omega))$ and $\Delta = (t_0 = s_0 < s_1 < \dots < s_r = t_1)$ a subdivision of the internal $[t_0, t_1]$. By connecting $Y(t_{i-1}, \omega)$ with $Y(t_i,\omega)$ by a geodesic curve for $i = 1,2,\dots,n$, we obtain a piece-wise smooth curve $C_\Delta(\omega)$ approximating $C(\omega)$. Take a random tensor $U^\Delta_2(\omega) \in T^n_m(Y(t_1,\omega))$ parallel to $U_1(\omega)$ along $C_\Delta(\omega)$. Then the random tensor in $T^n_m(Y(t_1,\omega))$:

$$U_2(\omega) = \underset{|\Delta|\to 0}{\text{l.i.p.}}\ U^\Delta_2(\omega)\ , \qquad |\Delta| = \max_i\ (t_i - t_{i-1}),$$

is said to be parallel to $U_1(\omega)$ along $C(\omega)$.

We will give two interesting examples of application of stochastic parallelism.

1. <u>Diffusions of tensors</u>. The diffusion of tensors was introduced by K. Itô [2] for the diffusion of covariant tensors on a Riemannian manifold induced by the Brownian motion on the manifold and extended to by E.B. Dynkin [3] to the general case we are dealing with below. Let $\{x_t(\omega)\}$ be a diffusion on S with generator

$$(6) \qquad \mathcal{J} = 1/2a^{ij}\ \nabla_i\nabla_j \neq b^i\nabla_i.$$

If $u(x)$ is a scaler field, we define a semi-group

$$H_t u(a) = E_a(u(X_t)) \tag{7}$$

and we have

$$\lim_{t \downarrow 0} \frac{H_t u(a) - u(a)}{t} = (1/2 a^{ij} \nabla_i \nabla_j + b^i \nabla_i) u(a) \quad \text{for } u \text{ smooth.} \tag{8}$$

If $u(x)$ is a tensor field of type (m,n), we have

$$u(X_t) \in T^n_m(X_t) \quad \text{but} \quad u(X_t) \notin T^n_m(X_0) \equiv T^n_m(a),$$

so the right hand side of (7) has no meaning. Hence we interpret (3) as follows. Take a tensor $\hat{U}_t(\omega)$ in $T^n_m(a)$ $(= T^n_m(X_0))$ such that $u(X_t)$ is parallel to $\hat{U}_t(\omega)$ along the curve $\{X_t(\omega)\}$. Then

$$H_t u(a) = E_a(\hat{U}_t(\omega)) \tag{7'}$$

defines a semi-group with generator $\mathcal{G}$ of the form (6). The family of tensors $V_t(\omega) \in T^m_n(X_t)$ parallel to $v(a) \equiv v(X_0)$ along the curve $(X_t(\omega))$ is a diffusion on the bundle T^m_n defined by Dynkin [3]. Since $(X_t(\omega))$ is determined by a stochastic differential equation:

$$dX^i_t = \sigma^i_\alpha \, d\xi^\alpha + \rho^i \, dt, \tag{9}$$

where $(\xi_t(\omega))$ is a Brownian motion and

$$a^{ij} = \sum_\alpha \sigma^i_\alpha \sigma^j_\alpha \quad \text{and} \quad b^i = \rho^i + a^{jk} \Gamma^i_{jk} \quad ,$$

$V_t(\omega)$ is determined by the stochastic differential equation:

$$dV_t = (\Gamma_i V_t) \circ dX^i_t \; , \tag{10}$$

i.e.

$$dV_t = (\Gamma_i V_t) \circ (\sigma^i_\alpha \, d\xi^\alpha + \rho^i \, dt).$$

To obtain (7') it is crucial to note that the equation

$$(11) \qquad (V_t, u(X_t)) = (v(a), \hat{U}_t(\omega))$$

holds by virtue of the invariance of the bilinear form (1) under parallel displacement.

2. Rolling along Brownian motion.

Let S and $\tilde{S}$ be ℓ-dimensional Riemannian manifolds with metric tensors $\{g_{ij}\}$ and $\{\tilde{g}_{ij}\}$ respectively. Let

$$C: x(t), \; t_0 \leq t \leq t_2.$$

be a curve on S. If we roll $\tilde{S}$ on S along C without slipping, the touching point on $\tilde{S}$ describes a curve

$$\tilde{C}: \tilde{x}(t), \; t_0 \leq t \leq t_2.$$

Let $\{v_\alpha(t)\}_{\alpha=1,2,\ldots,\ell}$ denote the family of covariant vectors in $T_1(x(t))$ obtained from an orthonormal basis in $T_1(x(t_0))$ by parallel displacement along C. Then $\{v_\alpha(t)\}$ is an orthonormal basis in $T_1(x(t))$ for every t. Similarly we obtain a family of covariant vectors $\{\tilde{v}_\alpha(t)\}_{\alpha=1,2,\ldots,\ell}$ in $\tilde{T}_1(\tilde{x}(t))$ by parallel displacement along $\tilde{C}$. $\{\tilde{v}_\alpha(t)\}$ is also an orthonormal basis in $\tilde{T}_1(x(t))$, as we can prove by using the symmetric chain rule (C_s). We choose the initial orthonormal basis $\{\tilde{v}_\alpha(t_0)\}$ so that $\{\tilde{v}_\alpha(t_0)\}$ coincides with $\{v_\alpha(t_0)\}$ when S starts rolling. As S rolls, $\tilde{x}(t)$ coincides with $x(t)$, $\tilde{T}_1(\tilde{x}(t))$ with $T_1(x(t))$, $\{\tilde{v}_\alpha(t)\}$ with $\{v_\alpha(t)\}$ and $d\tilde{x}(t)/dt$ with $dx(t)/dt$ at every t. Thus we have

$$(12) \qquad \tilde{v}_{\alpha,i} \frac{d\tilde{x}^i}{dt} = v_{\alpha,i} \frac{dx^i}{dt}$$

$$(13) \qquad \frac{dv_{\alpha,i}}{dt} = \Gamma^k_{ji} \, v_{\alpha,k} \frac{dx^j}{dt}$$

$$(14) \qquad \frac{d\tilde{v}_{\alpha,i}}{dt} = \tilde{\Gamma}^k_{ji} \, \tilde{v}_{\alpha,k} \frac{d\tilde{x}^j}{dt}$$

Since $x(t)$ is given, $v_{\alpha,i}(t)$ is determined by (13). By solving (12) and (14) for $\tilde{x}^i(t)$ and $\tilde{v}_{\alpha,i}(t)$, we obtain the curve $\tilde{C}$: $\tilde{x}^i(t)$, $t_0 \leq t \leq t_1$.

Let $C(\omega)$: $\{x_t(\omega)\} t \in [0,\infty)$ be Brownian motion on S. Since the sample curve of $C(\omega)$ is not smooth, we cannot apply the above argument to each sample curve to obtain the random curve $\tilde{C}(\omega)$ on $\tilde{S}$ by rolling $\tilde{S}$ on S along $C(\omega)$. However, the geodesic approximation method will work as for stochastic parallelism. Then $\tilde{C}(\omega)$: $\{X_t(\omega)\} t \in [0,\infty)$ is determined by the following stochastic differential equations analogous to (12), (13) and (14):

(12') $$\tilde{V}_{\alpha,i} \circ d\tilde{X}^i = V_{\alpha,i} \circ dX^i$$

(13') $$dV_{\alpha,i} = \Gamma^k_{ji} V_{\alpha,k}) \circ dX^j$$

(14') $$d\tilde{V}_{\alpha,i} = (\tilde{\Gamma}^k_{ji} \tilde{V}_{\alpha,k}) \circ d\tilde{X}^j ,$$

from which we can prove that $\{\tilde{X}_t(\omega)\}$ is a diffusion on S with generator

(15) $$\tilde{\mathcal{G}} = 1/2\, \tilde{g}^{ij} \tilde{\nabla}_i \tilde{\nabla}_j ,$$

i.e. Brownian motion on S. Using the symmetric chain rule (C_s), we can easily verify the formula (15) in terms of the geodesic coordinates on S and $\tilde{S}$ at the starting points of C and $\tilde{C}$. The rolling problem along Brownian motion was first discussed by H.P. McKean [4] in the case that S_1 is a plane and S_2 is a sphere.

Bibliography

[1] K. Itô, Stochastic differentials, to appear in Applied Mathematics and optimization.

[2] K. Itô, The Brownian motion and tensor fields on Riemannian manifold, Proc. Int. Congress Math. 1962 (Stockholm), 536-539.

[3] E.B. Dynkin, Diffusions of tensors, Soviet Math. Dokl. Vol. 9 (1968), No. 2, 532-535.

[4] H.P. McKean, Jr., Brownian motions on the 3-dimensional rotation group, Mem. Coll. Sci., Univ. Kyoto, Ser. A, 33, Math. No. 1 (1960), 25-38.

[5] R.L. Stratonovich, Conditional Markov processes and their application to optimal control, Elsevier, New York (1968).

[6] D.L. Fisk, Quasi-martingales and stochastic integrals, Technical Report No. 1, Dept. of Math., Michigan State University, 1963.

Diffusion processes in bounded domains and singular perturbation problems for variational inequalities with Neumann boundary conditions

A. Bensoussan and J. L. Lions

Introduction

We have observed in a recent paper (Bensoussan-Lions [1]) that problems of optimal stopping times for stochastic or deterministic systems in $\mathbb{R}^n$ are equivalent to variational inequalities (V.I.) of evolution or of stationery type, for second order parabolic or elliptic operators in the stochastic case and for 1st order hyperbolic operators in the deterministic case. (One can see references to variational inequalities in the paper [1] of the authors.) We also mention the equivalence of some of the V.I. which are met in this framework with free boundary problems of the Stefan's type.

When the stochastic (Itô's) state operation converges (as the variance tends to zero) to a deterministic ordinary differential equation, it then follows from the interpretation of the solution of the V.I. as optimal stopping times problems that the solution of the parabolic (or elliptic) V.I. converges to the solution of the hyperbolic V.I.; this is a singular perturbation result for V.I.

We would like to thank very much N. El Karoui and S. Varadhan for interesting discussions and suggestions.

The next step is to extend the above remark to the case of diffusion processes defined in bounded domains, say in $\mathcal{O} \subset \mathbb{R}^n$. One can then introduce a number of different situations arising on the boundary $\partial\mathcal{O}$ of $\mathcal{O}$; these situations give rise to different boundary value problems for V.I. and, at least formally, it is clear that in this way we shall obtain a number of results for singular perturbations of V.I. with different boundary conditions.

This program was made precise in Beusoussan-Lions [2] for the case of diffusion with absorption at the boundary--which leads to the Dirichlet boundary conditions.

We continue here the more complicated case of Neumann boundary conditions.

We present a new approach to the problem of the strong formulation of diffusion processes in bounded domains, very much in the spirit of the usual approach for stochastic differential equations in R^n. We make Lipschitz assumptions on the drift and the variance, and we construct the solution using Picard's iteration technique, as it is done in R^n. Hence we do not need to consider first a weak formulation of the problem, as it is usually done in the literature.

We study next the limit of the process when the variance term goes to 0 and we prove that it converges towards the solution of a differential equation with projection of the velocity on the boundary of the domain.

We then apply this model to obtain the desired singular perturbation results. We confine ourselves to the stationary case, but the whole theory extends to the evolution case.

1. Assumptions and Notation

Let $(\Omega, \mathscr{A}, P)$ be a probability space and let $\mathscr{F}_t$ be an increasing family of sub σ algebras of $\mathscr{A}$; let $w(t)$ be a standard n dimensional Wiener process with respect to $\mathscr{F}_t$. Let x_0 be a random variable with values in R^n and ξ_0 be a scalar random variable such that

(1.1) x_0, ξ_0 are $\mathscr{F}_0$ measurable and $x_0 \in \bar{\mathcal{O}}$,

(1.2) $\bar{\mathcal{O}}$ is a bounded domain of R^n whose boundary is a C^3 manifold.

We next consider two functions $g(x,t)$ and $\sigma(x,t)$ from $R^n \times [0,\infty) \to R^n$ and $R^n \times [0,\infty) \to \mathscr{L}(R^n;R^n)$ such that

(1.3) $|g(x,t)-g(x',t)| + |\sigma(x,t)-\sigma(x',t)| \leq C|x-x'|$

(1.4) $|g(x,t)|, \ |\sigma(x,t)| \leq C.$

On $\Gamma = \partial\mathcal{O}$, we consider a vector field $\gamma(x)$, $x \in \Gamma$ such that

(1.5) γ is continuous from $\Gamma \to R^n$ and $\gamma(x)\cdot n(x) \geq \delta > 0$, $\forall\, x \in \Gamma$, where $n(x)$ is the outward unit normal in x.

We define the following problem: find a process $y(t)$ with values in R^n such that

(1.6) $y(t)$ is adapted to $\mathscr{F}_t$ and continuous,

(1.7) there exists a scalar process $\xi(t)$ which is continuous, non decreasing, adapted, such that $y(t)$ has the stochastic differential

$$dy(t) = g(y(t),t)dt + \sigma(y(t),t)dw(t) - \chi_\Gamma(y(t))\gamma(y(t))d\xi(t),$$

(1.8) a.s. $y(t) \in \bar{\mathcal{O}} \quad \forall t$,

(1.9) a.s. $\int_{t_1}^{t_2} \chi_{\mathcal{O}}(y(t))\,d\xi(t) = 0, \forall\, t_1 \le t_2$,

(1.10) $y(0) = x_0$, $\xi(0) = \xi_0$.

Such a process $y(t)$ will be called <u>a diffusion process with reflection on the boundary of</u> $\mathcal{O}$. Usual formulations of the problem (weak sense) allow the probability measure P and the process $w(t)$ to be chosen besides the pair $y(t)$, $\xi(t)$, such that for this P, $w(t)$ is a Wiener process and (1.6),...,(1.10) hold. The existence and uniqueness (in law) can then be obtained by the submartingale approach of STROOCK-VARADHAN [7], or by the methods of EL KAROUI [4] and WATANABE [8]. It is then possible to weaken the assumptions on g and σ and assume only continuity instead of Lipschitz properties. Here we are looking for a strong existence result. However, to derive strong existence from weak existence it is enough to prove strong uniqueness(*). But this would be a long detour and it is well known that in R^n one can get the strong existence result without proving first the weak existence. In our approach, we will not use the weak existence theory whatsoever.

2. Existence and uniqueness theorem

We shall prove the following:

<u>Theorem 2.1. Under the assumptions</u> (1.1), (1.2), (1.3), (1.4), (1.5) <u>there exists one and only one process</u> $y(t)$ <u>satisfying</u> (1.6),...,(1.10).

(*) This was pointed out to one of the authors by S. VARADHAN.

Proof: The proof works in two stages. We first consider the problem in a half plane and next use a system of local maps for describing the boundary Γ and a localization procedure.

In the half plane $\{x | x_n > 0\}$ we first solve an auxiliary problem: let $g(t)$ and $\sigma(t)$ be two adapted stochastic processes with values in R^n and $\mathcal{L}(R^n;R^n)$ respectively and let x_0 be a random variable with values in R^n such that $x_{0n} \geq 0$ and $E|x_0|^4 < \infty$, let ξ_0 be a scalar R.V. such that $E|\xi_0|^4 < \infty$, and let us assume that x_0, ξ_0 are $\mathcal{F}_0$ measurable. We set the following problem: find $y(t)$ adapted and continuous, $\xi(t)$ (scalar) non decreasing adapted and continuous such that

$$(2.1)\quad dy_i(t) = g_i(t)dt + \sum_{j=1}^{n} \sigma_{ij}(t)dw_j(t), \quad i = 1,\ldots,n-1$$

$$dy_n(t) = g_n(t)dt + \sum_{j=1}^{n} \sigma_{nj}(t)dw_j(t) + d\xi(t)$$

$$(2.2)\quad y_n(t) \geq 0$$

$$(2.3)\quad y_n(t)d\xi(t) = 0$$

$$(2.4)\quad y(0) = x_0, \ \xi(0) = \xi_0.$$

Let y^1, y^2 be two solutions. Clearly $y_i^1 = y_i^2$, $i = 1,\ldots,n-1$ and

$$(d\xi^1-d\xi^2)(\xi^1-\xi^2) = (d\xi^1-d\xi^2)(y^1-y^2) \leq 0$$

i.e.

$$|\xi^1(t)-\xi^2(t)|^2 \leq 0, \forall\ t$$

hence

$$\xi^1(t) = \xi^2(t), \ y^1(t) = y^2(t) \text{ a.s.}$$

For the existence, it is obvious that it is enough to work in one dimension. Hence we want a pair $y(t)$, $\xi(t)$ such that

(2.5) $y(t)$, $\xi(t)$ are continuous and adapted; $\xi(t)$ is non decreasing

(2.6) $dy = g\ dt + \sigma\ dw(t) + d\xi$

(2.7) $y(t) \geq 0$

(2.8) $y(t)d\xi(t) = 0$

(2.9) $y(0) = x_0$, $\xi(0) = \xi_0$.

It is also enough to consider an arbitrary finite interval of time $[0,T]$. We consider a discretization procedure, let $h = \frac{T}{N}$ $(N \to +\infty)$ and set

$$g_0 = 0$$

$$g_k = \frac{1}{h}\int_{(k-1)h}^{kh} g(t)\,dt,\ k \geq 1; \quad g^h = \begin{vmatrix} g_{k-1} & \text{in } [kh,(k+1)h) \\ 0 & \text{in } [0,h) \end{vmatrix}$$

$$f_k = \int_0^{hk} \sigma(t)\,dw(t);$$

we define a sequence of R.V. y_k by

$$y_0 = x_0$$

$$y_k = (y_{k-1}+hg_{k-1}+f_k-f_{k-1})^+.$$

By easy computations one proves the following estimates

(2.10) $E\ y_k^2 \leq C$

$$E(y_p-y_k)^2 \leq Ch(p-k)$$

$$E(y_p-y_k)^4 \leq C[(p-k)h]^{3/2}.$$

One next defines the process

$$(2.11)\quad y_h(t) = \begin{cases} x_0 & \text{in } [0,h] \\ \dfrac{(k+1)h-t}{h}\, y_{k-1} + \dfrac{(t-kh)\,y_k}{h} & \text{in } [kh,(k+1)h]. \end{cases}$$

From (2.10) it follows that

$$(2.12)\quad E(y_h(t)-y_h(s))^2 \le C(t-s)$$

$$E(y_h(t)-y_h(s))^4 \le C(t-s+o(h))^{3/2}$$

$$Ey_h(t)^2 \le C$$

$$E\, y_h(t)^4 \le C.$$

Using Ascoli's theorem it follows from (2.12) that one can extract a subsequence $y_{h'}$ such that

$$E\, y_{h'}(t)z \to E\, y(t)z \text{ in } C(0,T),\ \forall\ z \in L^2.$$

Also $y(t)$ is an adapted process and from (2.12) we get using Fatou's lemma

$$E(y(t)-y(s))^4 \le C(t-s)^{3/2}$$

which implies that $y(t)$ is a continuous process. Next from the recurrence relationship it follows that

$$(2.13)\quad y_h(t) - y_h(s) \ge \int_{(k_1+1)h}^{k_2h} g^h(t)\,dt + \int_{k_1h}^{(k_2-1)h} \sigma(t)\,dw(t)$$

$$+ \frac{(k_1+1)h-s}{h}\,(y_{k_1}-y_{k_1-1})$$

$$+ \frac{(t-k_2h)}{h}\,(y_{k_2}-y_{k_2-1})$$

where k_1, k_2 are such that

$$s \in [k_1 h, (k_1+1)h], \quad t \in [k_2 h, (k_2+1)h].$$

By passing to the limit in (2.13) one obtains

$$y(t) - y(s) \geq \int_s^t g(\lambda)\,d\lambda + \int_s^t \sigma(\lambda)\,dw(\lambda), \quad s \leq t,$$

and we may define $\xi(t)$ by

$$\xi(t) = y(t) - \int_0^t g(\lambda)\,d\lambda - \int_0^t \sigma(\lambda)\,dw(\lambda),$$

and the pair $y(t)$, $\xi(t)$ satisfies (2.5), (2.6), (2.7) and (2.9). To prove (2.8) one first proves, using the approximation procedure, that

$$E\, y_k^2 \leq E\, x_0^2 + 2E \int_h^{(k+1)h} \tilde{y}_h(t) g^h(t)\,dt + E \int_0^{kh} \sigma^2(t)\,dt$$

where

$$\tilde{y}_h(t) = \begin{cases} 0 & \text{on } [0,h) \\ y_{k-1} & \text{in } [kh,(k+1)h), \end{cases}$$

and by a limit argument one obtains:

$$E\, y(t)^2 \leq E\, x_0^2 + 2E \int_0^t y(s) g(s)\,ds + E \int_0^t \sigma^2 ds$$

which taking into account Itô's formula yields

$$E \int_0^t y(s)\,d\xi(s) \leq 0, \quad \forall\, t$$

hence (2.8) follows.

The auxiliary problem (2.1),...,(2.4) being solved, we can now consider our starting problem in a half plane. We want to find $y(t)$ continuous and adapted, with $E \int_0^T |y(t)|^4 dt < \infty$, and $\xi(t)$ continuous adapted and non decreasing such that

(2.14) $dy_i(t) = g_i(y,t)dt + \sum_{j=1}^{n} \sigma_{ij}(y,t)dw_j(t),$

$$i = 1,\ldots,n-1$$

(2.15) $dy_n(t) = g_n(y,t)dt + \sum_{j=1}^{m} \sigma_{nj}(y,t)dw_j(t) + d\xi(t)$

(2.16) a.s. $y_n(t) \geq 0, \ \forall\, t$

(2.17) $y_n(t)d\xi(t) = 0$

(2.18) $y(0) = x_0, \ \xi(0) = \xi_0.$

To prove the uniqueness we note that if y^1, y^2 are two solutions then

$$(y^2-y^1, d(y^2-y^1)) \leq (g(y^2)-g(y^1), y^2-y^1)dt$$
$$+((\sigma(y^2)-\sigma(y^1))dw, y^2-y^1)$$

hence by Itô's formula and by taking mathematical expectations

$$E|y^2(t)-y^1(t)|^2 \leq C\int_0^t E|y^2(s)-y^1(s)|^2 ds$$

which implies a.s. $y^1(t) = y^2(t)$, $\forall\, t$.

To prove existence, we consider an iteration procedure as follows: set

$$y_0(t) = x_0.$$

Having defined $y_p(t)$, define $y_{p+1}(t)$ by solving the auxiliary problem (2.1),...,(2.4) corresponding to

$$g(t) = g(y_p(t),t)$$

$$\sigma(t) = \sigma(y_p(t),t).$$

By computations very similar to the R^n case, one arrives at

$$E|y_{p+1}(t)-y_p(t)|^4 \leq C_T \int_0^t E|y_p(s)-y_{p-1}(s)|^4 ds$$

hence

$$E|y_{p+1}(t)-y_p(t)|^4 \leq \frac{C_T' T^p}{p!} .$$

By a standard arument using Borel Cantelli's lemma, it follows that

$$x_0 + \sum_{p=0}^{\infty} (y_{p+1}(t)-y_p(t))$$

converges a.s. uniformly in t towards a process y(t). If one defines

$$\begin{aligned} \xi(t) = \xi_0 + y_n(t) - x_{on} - \int_0^t g_n(y(s),s)\,ds \\ - \int_0^t \sum_j \sigma_{nj}(y(s),s)\,dw_j(s) \end{aligned}$$

one can easily check that y(t), $\xi(t)$ satisfy (2.14), (2.15), (2.16) and (2.18). To prove (2.17) one first notes that

$$\begin{aligned} \tfrac{1}{2}|y_p(t)|^2 = \tfrac{1}{2}|x_0|^2 + \int_0^t (y_p(s),g(y_{p-1}(s)))\,ds \\ + \int_0^t (y_p(s),\sigma(y_{p-1}(s))\,dw(s)) \\ + \frac{1}{2}\int_0^t \mathrm{tr}\; \sigma(y_{p-1}(s))\sigma^*(y_{p-1}(s))\,ds. \end{aligned}$$

Passing to the limit and comparing with Itô's formula, one easily gets (2.17).

Now to solve the problem in a domain $\mathcal{O}$ instead of R^n, one considers a covering $\mathcal{O}_1,\ldots,\mathcal{O}_N$ of $\Gamma = \partial$ by a system of local maps. One defines the process till the first time when it reaches the boundary of $\mathcal{O} \cup \mathcal{O}_1$.

Inside $\mathcal{O}$ the process is defined as a usual diffusion and in $\mathcal{O}_1$ as the image by a diffeomorphism of a diffusion with reflection on a half plane. By degrees, one defines the process in any interval $[0,T]$. For complete details see BENSOUSSAN-LIONS [3].

3. Deterministic case

We did not assume so far that σ was non degenerate In particular we can take $\sigma = 0$. The problem is then completely deterministic and one gets more precise results which can be obtained in a way which is completely analogous to the stochastic case (with of course many simplifications). We state the result without proof.

Theorem 3.1. We assume the same properties on $\mathcal{O}$(*) *and* g *as for theorem* 2.1. *Then there exist* $y(t)$ *continuous with* $\frac{dy}{dt} \in L^\infty(0,T;R^n)$ *and* $\eta(t)$ *measurable and* $L^\infty(0,T)$ *such that*

(3.1) $\eta(t) \geq 0$ a.e. t

(3.2) $y(t) \in \bar{\mathcal{O}}$, $\forall$ t

$$(3.3)\quad \begin{cases} \frac{dy}{dt} = g(y,t) - n(y)\chi_{\partial\mathcal{O}}(y)\eta \\ y(0) = x_0 \end{cases}$$

(3.4) $\eta(t)\chi_{\mathcal{O}}(y(t)) = 0$ a.e. t.

It is possible to completely eliminate $\eta(t)$ from (3.3) and (3.4) Defining

(3.5) $\Gamma^+(t) = \{x \in \Gamma \mid g(x,t) \cdot n(x) > 0\}$

(*) However, it is enough for Γ to be of class C^2.

and setting, still for $x \in \Gamma$

(3.6) $\pi g(x,t) = g(x,t) - n(x)(g(x,t),n(x))$

then $y(t)$ is a solution of (3.2) and

$$\frac{dy}{dt} = g(y,t)\chi_{\mathcal{O}}(y) + \pi g(y,t)\chi_{\Gamma^+(t)}(y) \tag{3.7}$$
$$y(0) = x_0.$$

It is useful, for the singular perturbation problems we will consider in the next paragraph, to consider the case when $\mathcal{O}$ is convex. In this case we can replace (3.7) by

$$\begin{cases} \left(\frac{dy}{dt} - g(y,t), z-y(t)\right) \geq 0 \;\; \forall\, z \in \bar{\mathcal{O}}, \quad \text{a.e. } t \\ y(0) = x_0 \end{cases} \tag{3.8}$$

still keeping (3.2).

Similar considerations can be applied to the stochastic case if $\gamma(x) = n(x)$. The process $y(t)$ obtained in theorem 2.1 can be characterized in a unique way as follows ($\mathcal{O}$ convex)

(3.9) $y(t)$ is continuous adapted, a.s. $y(t) \in \bar{\mathcal{O}}$, $\forall\, t$ and $y(t) - x_0 - \int_0^t g(y)\,ds - \int_0^t \sigma(y)\,dw(s)$ is a process with bounded variation, equal to 0 for $t = 0$.

(3.10) $(z(t)-y(t), dy(t)-g(y(t),t)dt-\sigma(y(t),t)dw(t)) \geq 0$ for any process $z(t)$ continuous adapted such that $z(t) \in \bar{\mathcal{O}}$, $\forall\, t$.

Problem (3.8) is a <u>variational inequality</u> and (3.9), (3.10) a <u>stochastic variational inequality</u>.

Let us now consider the case when $\sigma(x,t) = \epsilon I$, and let us denote by $y_x(t)$ the solution of (3.8) corresponding to $x_0 = x$, and $y_x^\epsilon(t)$ the solution of (3.9), (3.10) also corresponding to $x_0 = x$. We have the following convergence result

<u>Theorem 3.1.</u> <u>Let us assume that</u> $\mathcal{O}$ <u>is convex and that</u> $g(x,t) = g(x)$ <u>satisfies</u> $|g(x)-g(x')| \leq C|x-x'|$ <u>and let</u> $\alpha > 2C$; <u>then we have</u>

$$(3.11)\quad e^{-\alpha t}E|y_x^\epsilon(t)-y_x(t)|^2 \leq \frac{n\epsilon^2}{\alpha},$$

$$(3.12)\quad \int_0^\infty e^{-\alpha t}E|y_x^\epsilon(t)-y_x(t)|^2 dt \leq \frac{n\epsilon^2}{\alpha(\alpha-2c)}.$$

<u>Proof</u>: Setting $y_x'(t) = \frac{dy_x}{dt}(t)$, we can apply Itô's formula to $y_x^\epsilon(t) - y_x(t)$, which yields

$$(3.13)\quad e^{-\alpha t}|y_x^\epsilon(t)-y_x(t)|^2 = \int_0^t [-\alpha e^{-\alpha s}|y_x^\epsilon(s)-y_x(s)|^2 + n\epsilon^2 e^{-\alpha s}]ds + \int_0^t 2e^{-\alpha s}(y_x^\epsilon(s)-y_x(s), dy_x^\epsilon(s)-y_x'(s)ds).$$

From (3.8) and (3.10) it follows, by using suitable test functions ($z(t) = y_x(t)$ in (3.10) and $z = y_x^\epsilon(t)$ in (3.8)), that

$$(3.14)\quad \int_0^t e^{-\alpha s}(y_x^\epsilon(s)-y_x(s), -dy_x^\epsilon(s)+y_x'(s)ds + (g(y_x^\epsilon)-g(y_x))ds+\epsilon dw(s)) \geq 0.$$

Using (3.14) in (3.13) and taking mathematical expectations we get (taking into account the Lipschitz condition on g)

$$e^{-\alpha t}E|y_x^\epsilon(t)-y_x(t)|^2 + (\alpha-2C)\int_0^t e^{-\alpha s}E|y_x^\epsilon(s)-y_x(s)|^2 ds \le \frac{n\epsilon^2}{\alpha}$$

hence (3.11), (3.12).

The result of the preceding theorem can be extended to domains $\mathcal{O}$ and vector fields $\gamma(x)$ such that there exists a diffeomorphism $y = \psi(x)$, from $\mathcal{O} \to \tilde{\mathcal{O}}$, such that ψ and ψ^{-1} are C^3, $\tilde{\mathcal{O}}$ is convex and $\frac{\partial\psi}{\partial x} \cdot \gamma = \tilde{n}$, where $\tilde{n}$ is the outward normal on the boundary of .

In two dimensions, this is true for any regular domain without holes, as one can see, by using conformal mapping. In higher dimensions it is also true, provided $\mathcal{O}$ has "no holes" (in the sense "every surface inside can be retracted to a point"), provided the dimension is large enough; but this is a deep result in differential topology, due to Smale; its use here would seem rather strange.

4. Optimal stopping time problems

Let us first start with the deterministic case. We consider the solution of (3.7) with

(4.1) $g(x,t) = g(x)$, $|g(x)-g(x')| \le c|x-x'|$,

(4.2) $\mathcal{O}$ convex or can be transformed into a convex set by a diffeomorphism of class C^2 which transforms the normal on $\partial\mathcal{O}$ into the normal on the boundary of the image of $\mathcal{O}$,

(4.3) f is a functional on $\bar{\mathcal{O}}$, such that

$$|f(x)-f(x')| \le L|x-x'|, \forall\, x,x' \in \bar{\mathcal{O}}$$

(4.4) $\alpha > 2C$.

For simplicity we will only consider the case when $\mathcal{O}$ is convex. We define

$$(4.5) \quad u(x) = \operatorname*{Inf}_{s \geq 0} \int_0^s e^{-\alpha t} f(y_x(t))\,dt$$

where $y_x(t)$ has been previously defined (solution of (3.8) for $x_0 = x$).

We next have

<u>Theorem 4.1. Under assumptions</u> (4.1),...,(4.4) <u>and assuming</u> $\mathcal{O}$ <u>to be convex, the function</u> u <u>defined by</u> (4.5) <u>satisfies</u>

$$(4.6) \quad u \text{ is uniformly Lipschitz in } \bar{\mathcal{O}}, \quad u|_\Gamma \in W^{1,\infty}(\Gamma)$$

$$(4.7) \quad \begin{cases} u \leq 0, \ -\sum_i g_i \dfrac{\partial u}{\partial x_i} + \alpha u \leq f, \\[2ex] u[-\sum_i g_i \dfrac{\partial u}{\partial x_i} + \alpha u - f] = 0 \text{ a.e. in } \mathcal{O} \end{cases}$$

$$(4.8) \quad \begin{cases} u \leq 0 \text{ on } \Gamma, \ -\sum_i (\pi g)_i \dfrac{\partial u}{\partial x_i} + \alpha u \leq f \text{ on } \Gamma^+ \\[2ex] u[-\sum_i (\pi g)_i \dfrac{\partial u}{\partial x_i} + \alpha u - f] = 0 \text{ on } \Gamma^+ \end{cases}$$

<u>Proof</u>: Using the variational inequality (3.8) written for $x_0 = x$ and $x_0 = x'$ successively, and reasoning as for Theorem 3.1, one gets

$$\sup_{t \geq 0} e^{-\alpha t} |y_x(t) - y_{x'}(t)|^2 \leq |x - x'|^2$$

$$\int_0^\infty e^{-\alpha t} |y_x(t) - y_{x'}(t)|^2 dt \leq \frac{1}{\alpha - 2C} |x - x'|^2,$$

which yields the Lipschitz property of u as it is easy to check. By Rademacher's theorem u is a.e. differentiable in $\bar{\mathcal{O}}$ and by a standard dynamic programming argument one gets (4.7). On the boundary Γ^+ the notation

$\sum_i (\pi g)_i \frac{\partial u}{\partial x_i}$ means that

$$\frac{u(x+\lambda\pi g+o(\lambda))-u(x)}{\lambda} \to \pi g \cdot \frac{\partial u}{\partial x}$$

where $x + \lambda\pi g + o(\lambda) \in \Gamma$ for λ small enough. By dynamic programming we also get (4.8).

We proceed next to the stochastic case. We consider first the following Neumann boundary value problem: find u satisfying (Δ being the Laplacian)

$$(4.9) \quad u \in C^1(\bar{\mathcal{O}}), \ \Delta u \in L^2(\mathcal{O})$$

$$(4.10) \quad -\frac{\varepsilon^2}{2}\Delta u - \sum_i g_i \frac{\partial u}{\partial x_i} + \alpha u \leq f \text{ a.e. in}$$

$$(4.11) \quad u \leq 0 \text{ in } \bar{\mathcal{O}}$$

$$(4.12) \quad u[-\frac{\varepsilon^2}{2}\Delta u - \sum_i g_i \frac{\partial u}{\partial x_i} + \alpha u - f] = 0 \text{ a.e. in } \mathcal{O}$$

$$(4.13) \quad \left.\frac{\partial u}{\partial n}\right|_\Gamma = 0.$$

We denote by $y_x^\varepsilon(t)$ the solution of (1.6),...,(1.10) corresponding to

$$(4.14) \quad g(x,t) = g(x); \quad \sigma(x,t) = \varepsilon I$$

$$(4.15) \quad \gamma(x) = n(x), \ x_0 = x$$

We have

Theorem 4.2. Assume (4.1), (4.3), (4.14), (4.15); then there exists one and only one solution of (4.9),...,(4.13), called u_ε. Furthermore we have

$$(4.16) \quad u_\varepsilon(x) = \operatorname*{Inf}_{\tau \geq 0} \int_0^\tau e^{-\alpha t} f(y_x^\varepsilon(t))\,dt$$

where τ is any stopping time; there exists an optimal stopping time $\hat{\tau}_\varepsilon$ realizing the inf in (4.16), given by

$$(4.17) \quad \hat{\tau}_\varepsilon(x) = \inf_{t > 0} \{y_x^\varepsilon(t) \notin C\}, \quad x \in \bar{\mathcal{O}}$$

<u>where</u>

$$(4.18) \quad C = \{x \in \bar{\mathcal{O}} \mid u_\varepsilon(x) < 0\}.$$

<u>Proof</u>: To prove the existence and uniqueness of a solution of (4.9),...,(4.13) one uses standard techniques of variational inequalities (see J.L. LIONS-G.STAMPACCHIA [6]). The procedure uses penalty techniques, a priori estimates and passing to the limit (for details see J.L. LIONS [5]). The interpretation of the solution u_ε by (4.16) is obtained in a way similar to that of A. BENSOUSSAN-J.L. LIONS [1], [2], using Itô's formula for functions satisfying the regularity properties (4.9) (note that unlike the solution of PDE, u_ε can not be in general a C^2 function).

We can finally study the limit of u_ε when $\varepsilon \to 0$. We have

<u>Theorem 4.3. Under the assumptions of Theorems</u> 4.1 <u>and</u> 4.2, <u>we have</u>

$$(4.19) \quad |u_\varepsilon(x)-u(x)| \leq c\varepsilon, \forall\, x \in \bar{\mathcal{O}}$$

<u>Proof</u>: This is an easy consequence of theorem 3.1, the interpretation of u and u_ε by formulas (4.5) and (4.16), and the Lipschitz property of f.

<u>Remark</u>. Singular perturbation results for variational inequalities of similar type but with <u>Dirichlet</u> boundary conditions have been obtained by the authors in [2].

References

[1] A. Bensoussan - J. L. Lions, Problèmes de temps d'arrêt optimal et inequations variationnelles paraboliques, Applicable Analysis, 1973, vol. 3, pp. 267-294.

[2] A. Bensoussan - J. L. Lions, Problèmes de temps d'arrêt optimal et perturbations singulieres pour les inéquations variationnelles avec conditions de Dirichlet sur le bord, Congrès IRIA Théorie du Contrôle, proceedings to appear Springer Verlag Lecture Notes.

[3] A. Bensoussan - J. L. Lions, Book in preparation.

[4] N. El Karoui, These Paris 1971

[5] J. L. Lions, Quelques méthodes de résolution des problèmes aux limites non linéaires, Dunod, Paris, 1969.

[6] J. L. Lions, G. Stampacchia, Variational inequalities, Comm. Pure and Applied Math. XX (1967), 493-519.

[7] D. W. Stroock - S.R.S. Varadhan, Diffusion processes with boundary conditions, Comm. Pure and Applied Math., XXIV, (1971), 147-225.

[8] S. Watanabe, On stochastic differential equations for multidimensional diffusion processes with boundary conditions, J. Math. Kyoto Univ., 1971.

ELLIPTIC ESTIMATES AND DIFFUSIONS IN RIEMANNIAN GEOMETRY AND COMPLEX ANALYSIS

Paul Malliavin

We shall deal here with some quantitative estimates obtained recently; this survey has no intention of being complete in any respect.

Contents

§0. Introduction

The use of the heat equation in geometric problems is generally worked out either by L^2 estimates of the corresponding elliptic problems, or by the construction of a parametrix. Ito's stochastic integral can be considered as an "infinitesimal parametrix" approach, which is therefore directly tied to the infinitesimal geometric invariants. The stochastic approach has also the advantage to associate to the heat equation the trajectories of the corresponding process, trajectories which have an immediate geometric meaning; it is indeed possible to consider such trajectories as limits of smooth curves, and so to apply the formalism of differential geometry. Finally the asymptotic of the trajectories is an approach to global problems. We shall denote by M a complete Riemannian manifold, by Δ_M the Laplace Beltrami operator on M.

Then there exists on the space of continuous paths on M starting from x_o a canonic probability measure. We shall denote by $\Omega_{x_o}(M)$ this space with this measure, and by $x_\omega(t)$, $\omega \in \Omega_{x_o}(M)$ a generic path (cf. [4,7]).

§1. Comparison equations and a Sturm theory

The idea is to replace the study of Δ_M by the study on an ordinary differential equation, which, as a comparison equation, will give estimates for Δ_M. The process $x_\omega(t)$, $\omega \in \Omega_{x_o}$ will be projected by some real function f; then the motion of $Y_\omega(t) = f(x_\omega(t))$ is governed by an Ito stochastic integral. It will be compared to a diffusion on $\mathbb{R}$.

Some of the results so obtained could also be gotten by an elementary application of the maximum principle in a "radial" coordinate system.

a) [10] Suppose M non-compact. Denote by p a real exhaustion function on M (i.e. $p^{-1}(K)$ is compact for every K compact), by $\|\nabla p(x)\|$ the length of the gradient of p, by $a(x) = (\Delta p)\|\nabla p\|^{-2}$.

Introduce
$$a^+(\tau) = \sup_{p(x)=\tau} a(x)$$
$$a^-(\tau) = \inf_{p(x)=\tau} a(x)$$
and the two following ordinary differential equations on $\mathbb{R}$:

$$L^\pm = \frac{d^2}{d\tau^2} + a^\pm(\tau)\frac{d}{d\tau} \quad .$$

Denote by g_M the Green function of M, $g^\pm$ the Green function of $L^\pm$. For every x_o fixed, there exists c_1, c_2 two constants such that

$$c_2\, g^+(p(x_o),p(x)) < g_M(x_o,x) < c_1\, g^-(p(x_o),p(x)), \quad \text{when} \quad p(x) \longrightarrow +\infty .$$

b) [2] Using as projection function the geodesic distance, then estimates of the lowest strictly positive eigenvalue of Δ_M on a geodesic ball are obtained in terms of bounds for the sectional curvature. It is shown, as a by-product, that the cut locus of a point on a compact manifold of negative curvature is of codimension 1. The lifetime of the diffusion on a complete manifold is also discussed.

c) [13] Asymptotic of the trajectories of the diffusion on a surface simply connected and of curvature $< -h < 0$. In particular, angular convergence (i.e. the existence of a geodesic asymptot) is obtained.

§2. Mean value formulae for harmonic forms and vanishings

If $O(M)$ denotes the principal bundle of orthonormal frames, then to a differential form π on M is associated an $\mathcal{R}^s$-valued function f_π, equivariant under the action of $O(n)$. If $\square$ denotes the de Rham-Hodge operator on M, then there is the Weitzenböck formula

$$f_{\square\pi} = -\Delta_{O(M)} f_\pi + J f_\pi$$

where $\Delta_{O(M)}$ is the horizontal Laplacian of Bochner and J a map of $O(M)$ into End $(\mathcal{R}^s)$. The harmonic forms are so lifted on $O(M)$ as the functions u satisfying the elliptic systems

2.1 $$\Delta_{O(M)} u - Ju = 0 \ .$$

a) [11] The elliptic system 2.1 can be integrated using the process associated to $\Delta_{O(M)}$ by a generalization of the Kac multiplicative functional method. We get, for all $t > 0$,

2.2 $$f_\pi(\tau_o) = E_{\tau_o}(R_\omega(t) f_\pi(\tau_\omega(t))), \qquad \omega \in \Omega_{\tau_o}(O(M))$$

where $R_\omega(t)$ is defined by $\frac{dR_\omega}{dt} = R_\omega(t)\, J(\tau_\omega(t))$, $R_\omega(0) = \text{Identity}$.

b) Using 2.2 and a maximum principle type argument, the vanishing theorem can be obtained in the case where the classical hypothesis of Bochner (some "positivity") is replaced by "positivity in the mean." This result depends [11] on a perturbation estimate.

c) [1] Integral representations for solutions of Spencer-Neumann boundary value problems for harmonic forms are obtained.

d) [18] Using hypoelliptic estimates, the infinitesimal method of Bochner leads to some new vanishings.

e) [19] Infinitesimal interpretations of the Weitzenböck formulae are worked out.

§3. Compound processes, multidimensional time processes

For an overdetermined elliptic system a diffusion associated to an elliptic operator of the system does not give a complete grasp of the behaviour of a solution of the system. To remedy this one can think of two approaches. One is to deal with a compound process for which the trajectories are locally defined by some operator of the system; the other is a multidimensional time process which includes all "possible" compound processes.

a) In the Siegel upper half plane

$$H = \{z \in \mathbb{C}^3;\ y_1 > 0,\ y_1^2 > y_2^2 + y_3^2\}$$

denote by Δ_1 the Laplacian associated to the Bergmann metric $\partial\bar\partial \log(y_1^2 - y_2^2 - y_3^2)$. Then the Fustenberg boundary on which bounded Δ_1-harmonic functions can be represented is essentially $\mathbb{R}^3 \times T$. Denote

$$\Delta_2 = 4y_1^2 \sum_{k\geq 1} \partial_k\bar\partial_k + 4 \sum_{k>1} y_1 y_k(\partial_1\bar\partial_k + \bar\partial_1\partial_k)$$

(Δ_2 is a linear combination of Hua operators.) Then, by a compound diffusion, it is proved that if $\Delta_1 f = \Delta_2 f = 0$, f bounded, then f is represented on the Shilov boundary $\mathbb{R}^3$. This approach depends on delicate numerical estimates.

b) [22] For the bidisk (z_1, z_2) a bi-dimensional time process can be constructed and an energy integral for bi-martingale associated to the following gradient of a bi-harmonic function f:

$$|\nabla_{12}f|^2 = |\partial_{12}f|^2 + |\partial_{1\bar{2}}f|^2 + |\partial_{\bar{1}2}f|^2 + |\partial_{\bar{1}\bar{2}}f|^2 \ .$$

§4. Diffusions on a semi-simple Lie group

On a real non-compact semi-simple Lie group with maximal compact subgroup K, the horizontal Laplacian Δ_G denotes the hypoelliptic operator $\Delta_G = \omega_G - \omega_K$ (where ω_G, ω_K are the Casimir operators); it induces a horizontal diffusion $g_\omega(t)$ on G which is the lifted diffusion on G of the Brownian on the Riemannian manifold G/K.

a) Asymptotic of $g_\omega(t)$. ([8]) In the Iwasawa decomposition NAK,

$$g_\omega(t) = n_{\omega_2}(t)\ \exp(a_{\omega_1}(t))k_{\omega_2}^{-1}(t)$$

where a_{ω_1} is a diffusion independent of the component ω_2 and where, when ω_1 is known, $(n_{\omega_2}(t), k_{\omega_2}(t))$ is a diffusion on the group product $N \times K$, with time dependent left invariant infinitesimal generator. Furthermore, when $t \longrightarrow +\infty$ almost surely

$$\text{limit}(n_{\omega_2}(t)) \quad \text{exists,}$$

$$a_{\omega_1}(t) = t\sum_{\alpha>0} Q_\alpha + 0(t^{\frac{1}{2}+\varepsilon}) \ ,$$

$$k_{\omega_2}(t) \quad \text{is ergodic on } K.$$

b) The lower bound $r_s(p)$ of the point spectrum of Δ_G on right p-K-equivariant function is a basic estimate for vanishing of L^s-bundle valued cohomology [20].

c) Estimates on $r_s(p)$ can be gotten either from an infinitesimal point of view using hypoelliptic estimates [9], or from a more global approach using [21] ergodic properties of $k_{\omega_2}(t)$.

§5. Lifted diffusion through a connection

a) [3] In local coordinate on M the Ito stochastic differential system associated to Δ_M can be written

5.1 $$dx^k = a_i^k\, db^i + c^k\, dt$$

where db^i are the differentials of the Wiener process on $\mathcal{R}$. Let u be a C^∞ function with support contained in $[0,1]$, and which has 1 for integral. Set

$$f_\varepsilon^i(t) = \int_0^\varepsilon b^i(t+\xi)\, u(\tfrac{\xi}{\varepsilon})\, d\xi \, .$$

Then the $b_\varepsilon^i(t)$ are C^∞ functions. Then consider the system of ordinary differential equations

5.2 $$dx_\varepsilon^k = a_i^k\, db_\varepsilon^i + [c^k - \frac{1}{2} \sum_{i,s} a_i^s\, \partial_s\, a_i^k]\, dt \, .$$

Then, $\varepsilon \longrightarrow 0$, the trajectories of 5.2 converge to the trajectories of 5.1. Now, given a construction in differential geometry which can be made on smooth curves, it could be applied to the solution of 5.2, and then, passing to the limit this construction is therefore applicable to 5.1. In terms of differential operators we get a mechanism to get from an elliptic operator another elliptic operator.

For instance, let F be a Riemannian vector bundle over M, and let us choose a connection on F. Then lifting the solution of 5.2 through the connection we get a natural lifting of the diffusion on M to a diffusion on F. The projection $F \longrightarrow M$ defines then a one-to-one correspondence between the trajectories of the lifted diffusion and those of the original diffusion.

b) [1] On the bundle of linear frame of a Riemannian manifold the de Rham-Hodge operator is associated to a diffusion with a vertical drift and no zero order term. The same can be used on tangent bundles.

§6. Elliptic regularity, fine harmonic functions, function algebras

The fine topology is the weakest topology for which superharmonic functions are continuous. If K is a compact set without interior for the usual topology, then K^f will denote its interior for the fine topology.

a) H(K) are the functions which are uniform limits on K on usual harmonic functions on some neighborhood of K. Then [15] if $K^f = \emptyset$, $H(K) = C(K)$. If $K^f \neq \emptyset$, the functions of H(K) are finely harmonic in the Fuglede sense. Furthermore, any $f \in H(K)$ is C^∞ in finely local L^2.

b) If $K \subset \mathbb{C}$, $K^f \neq 0$, let $\mathcal{H}(K) = \{f \in H(K); \bar{\partial} f = 0\}$. Then [16] every $z_0 \in K^f$ defines a continuous derivation on $\mathcal{H}(K)$. Gleason part of R(K) and fine connected component are related.

§7. Complex analysis

a) On a Stein manifold estimates of Green functions are obtained [14] in terms of an exhaustion function.

b) Poisson formulae for weakly pseudo-convex domains and analytic polyhedrons are derived [17] from the study of the exit measure of an adapted Kählerian diffusion.

August 1974

Bibilography

1. H. Airault: Approache stochastique a des problèmes aux limites pour les formes harmoniques (in preparation)

2. A. Debiard, J. Gaveau et E. Mazet: Temps d'arrêt des diffusions Riemanniennes, Comptes Rendus Acad. Sciences de Paris. 278,(1974), 723-725, 795-798.

3. P. Malliavin: Stochastic Geometry on Riemannian vector bundles, Journal of Functional Analysis, 1975.

4. I. Gihman and A. Skorohod, Stochastic Differential Equations, Ergebnisse der Mathematik, Bd. 72, Springer Verlag, New York, 1972.

5. K. Itô: The Brownian motion and tensor fields on Riemannian manifold, Proc. International Congress of Mathematicians, Stockholm, 1962, 536-639. See also this volume of Lecture Notes in Mathematics.

6. A. Koranyi and P. Malliavin: Poisson formula and compound diffusion associated to an overdetermined elliptic system on the Siegel halfplane of rank two, to appear.

7. H. P. McKean, Jr.: Stochastic Integrals, Academic Press New York, 1969.

8. M. P. Malliavin and P. Malliavin: Factorisation et lois limites de la diffusion horizontales au dessus d'un espace Riemannien symmetrique, Comptes Rendus de la Conference d'Analyse Harmonique et Théorie de Potentiel, Strasbourg 1973, Springer Verlag Lecture Notes in Mathematics, 404.

9. M. P. Malliavin et P. Malliavin: Spectre de l'operateur de Casimir d'un groupe de Lie semi-simple et théoremes d'annulation, Comptes Rendus Acad. Sciences de Paris, August 1974, 279, 185-188.

10. P. Malliavin, Asymptotics of the Green's function of a Riemannian manifold and Itô's stochastic integral, Proc. Nat. Acad. Sci. U.S.A. 71, 1974, 381-383.

11. P. Malliavin, Formules de la moyenne, Calcul de perturbations et théoremes d'annulation pour les formes harmoniques, Journal of Functional Analysis, 17, 1974, 274-291.

12. A. Milgram and P. Rosenbloom: Harmonic forms and heat conduction, Proc. Nat. Acad. Sci. U.S.A., 37, 1951, 180-184 and 435-438.

13. J. J. Prat: Etude asymptotique du mouvement Brownien sur une variété Riemannienne à courbure negative, Comptes Rendus Acad. Sciences de Paris, 272,1971, 1586-1589.

14. J. Vauthier, L^p estimés sur une variété de Stein, Comptes Rendus Acad. Sciences de Paris, 279, 1974, 409-

15. A. Debiard et B. Gaveau: Potentiel fin et algèbres de fonctions analytiques, Comptes Rendus Acad. Sciences de Paris, 278, 1974, 1025-1028.

16. A. Debiard et B. Gaveau: Potentiel fin et algèbres de fonctions analytiques I, Journal of Functional Analysis, 16, 1974, 289-304; II in same journal, 17, 1974,296-310.

17. A Debiard et B. Gaveau: Demonstration d'une conjecture de H. Bremmerman sur la frontiere de Shilov d'un domaine faiblement pseudo-convexe, Comptes Rendus Acad. Sciences de Paris 279, 1974, 407-

18. Mme. Latremolieres: Estimés hypoelliptiques et théoremes d'annulation Comptes Rendus Acad. Sciences de Paris, 279, 1974, 413

19. A. Lévy-Bruhl: Invariants infinitesimaux, Comptes Rendus Acad. Sciences de Paris 279, 1974, 197-200.

20. M. P. Malliavin et P. Malliavin: Diagonalisation de systèmes de de Rham-Hodge au dessus d'un espace Riemannien homogène, Conference d'Analyse Harmonique non Commutative, Luminy, Juiliet 1974, Springer-Verlag Lecture Notes in Mathematics (to appear 1975).

21. M. P. Malliavin et P. Malliavin: Proprietes ergodiques de la diffusion horizontale et spectre du Casimir d'un groupe semi-simple. (in preparation).

22. P. Malliavin: Processus à temps bidimensionnel dans le bidisque, Comptes Rendus Acad. Sciences de Paris, December 1974.

STOCHASTIC DIFFERENTIALS AND QUASI-STANDARD RANDOM VARIABLES

P. Greenwood, R. Hersh

Introduction

A "quasi-standard random variable" is obtained from a family of ordinary random variables depending on a parameter, by giving the parameter a non-standard value--for example, an infinitely large natural number, or a positive infinitesimal. Such a procedure corresponds, roughly, to letting n go to infinity, or letting h go to zero, where n and h respectively are natural numbers or positive real numbers. The advantage is that the quasi-standard random variable exists as a well-defined non-standard object, even if the corresponding limit of standard objects does not exist.

For example, if $\{X_i\}$ is a sequence of ordinary random variables and their averages are

$$A_n = \frac{1}{n} \sum_{i=1}^{n} X_i ,$$

then, if w is an infinite natural number,

$$A_w = \frac{1}{w} \sum_{i=1}^{w} X_i$$

is a quasi-standard random variable. If $\lim_{n\to\infty} A_n$ exists (in some sense), then the limit is (in some suitable sense) infinitely close to A_w. But in any case A_w exists, and possesses the properties it "should" have in common with the A_n ; for instance its distribution function satisfies any formulas which are satisfied by the distribution functions of A_n .

If f is any standard function defined on a real interval, and dt is an infinitesimal, then $df = f(t+dt) - f(t)$ is a well-defined quasi-standard function. Only if f is differentiable is df meaningful in standard analysis. Thus an important heuristic formula for the differential of Brownian motion,

(1) $$db^2 = dt$$

is meaningless in standard analysis, unless it is given a global interpretation by integrating both sides of the formula against some test function. In nonstandard analysis, (1) is meaningful locally (pointwise) and must be either true or false. In fact, it turns out to be false, as we will see below. We will also see, using nonstandard stochastic analysis, why the false formula (1) leads to correct formulas when summed or integrated.

In a recent paper [9], Professor Ito has shown how to define stochastic differentials as equivalence classes of interval functions. The procedure of the present note can be thought of as defining a representative for Ito's equivalence class, by choosing from the intersection of all intervals of the class. Only in the non-standard universe can this be done, for while the intersection contains plenty of quasi-standard infinitesimal intervals, it contains no standard intervals (since the intersection of all open intervals sharing a common left endpoint is empty.)

It would be interesting to develop a detailed comparison of the two methods of defining stochastic differentials, but we do not undertake this task here. Our objective is to show how quasi-standard random variables can be used in a natural and convenient manner in various calculations about continuous-time stochastic processes. Often an argument can be shortened, and burdensome technicalities avoided. In particular, it commonly happens in standard analysis that the obvious, "intuitive" way to go about things is to use the derivative, or perhaps a finite Taylor expansion; but this is forbidden, if we are dealing with Brownian motion, or some other non-differentiable process. The quasi-standard differential precisely follows that heuristic argument, which in standard analysis, has to be circumvented in some more or less roundabout way. We touch on questions of Holder continuity, quadratic variation, a version of Ito's lemma, and the martingale convergence theorem.

Quasi-standard random variables

The arguments and constructions we need from nonstandard analysis can be summed up in three basic principles: an Existence Principle, a Transfer Principle, and a Convergence Principle.

Existence Principle: The standard set-theoretic universe of ordinary analysis can be imbedded in a larger, non-standard universe, which contains, in particular, infinite natural numbers--i.e., nonstandard "natural numbers" which are larger than any standard positive natural number--and also infinitesimal positive numbers -- i.e., nonstandard "real numbers" which are greater than zero, but smaller than any standard positive real number. We write $x \sim y$ if $x - y$ is infinitesimal. If I is any set in the standard universe, $^*I$ denotes its nonstandard enlargement. $^*R$ is the enlargement of R, the set of standard real numbers; $^*N$ is the enlargement of N, the set of standard natural numbers. If x is a member of $^*R$ and x is not infinite, (i.e., there exists a standard real number c such that $|x| < c$) then there exists a unique standard real number $^\circ x$ such that $^\circ x \sim x$. This number is called the standard part of x. In particular, x is infinitesimal iff $^\circ x = 0$. If $f(x)$ is a function which maps $^*R$ into $^*R$ and whose values are finite, then $^\circ f$, when restricted to R, is a standard real-valued function.

Transfer Principle: Let $\mathcal{L}$ be the formal language of standard set theory furnished with enough constant symbols to provide a proper name for each of the standard numbers, sets and functions of standard analysis.

A sentence of $\mathcal{L}$ is true in the standard universe if and only if it is true, under an appropriate interpretation, in the non-standard universe. In particular, if a true sentence of $\mathcal{L}$ has the form "$\forall n \in N$, ... " then, in the non-standard interpretation, the sentence is true "$\forall n \in {}^*N$ ... " -- that is, for both finite and infinite natural numbers n. Similarly, if, for some standard positive real number a, a true sentence in $\mathcal{L}$ has the form "$\forall h \in R$, $0 < h \leq a$, ... " then in the non-standard interpretation, it is true "$\forall h \in {}^*R$, $0 < h \leq a$ ... " ; in particular, it is true if $h = dt$, where dt is a positive infinitesimal. We refer to a function as "standard" if it maps standard points into standard points. However, a standard function, when regarded as an object in the non-standard universe, will have non-standard points in its domain, and it may well map these into non-standard points. Thus the standard function x^2 takes on positive infinite values, if x is positive or negative infinite, and positive infinitesimal values, if x is positive or negative infinitesimal.

Convergence Principle: If $\{x_n\}$ is a standard sequence of real numbers, it converges to x if and only if $x_w \sim x$ for all infinite natural numbers w. It is bounded iff x_w is finite for all infinite w .

The Convergence Principle is Theorems 3.3.7 and 3.3.3 of Robinson [18]. The statements given here for the Existence Principle and the Transfer Principle are imprecise and unnecessarily restrictive; however, they are sufficient for our needs. For a more careful explanation, see [15] or [18], for example.

On the basis of the Existence Principle and the Transfer Principle, we can now formalize the definition of quasi-standard random variable.

Definition: Let I be a standard infinite set (e.g., N , the natural numbers, or R_+ , the positive reals, or $R_+ \times (0,1)$, the Cartesian product of R_+ and the open unit interval). For each i in I, let $z(i)$ be a standard random variable on some probability space. If k is a nonstandard "member" of I -- that is, a member of ${}^*I$ but not a standard member of I --then $z(k)$ will be called a quasi-standard random variable.

It is important to point out that there are other kinds of non-standard random variables. The quasi-standard ones are sufficient for our purposes, and they are particularly tractable for computations. A non-standard random variable, in general, would be simply an arbitrary function on a non-standard probability space, with values in ${}^*R$, measurable with respect to a non-standard probability measure. An important special class of non-standard random variables would be the "internal" random variables. In the terminology of nonstandard analysis, internal non-standard functions are those for which a **Transfer Principle** (suitably generalized from the very restricted version presented here) is valid. Our quasi-standard random variables are a particularly useful kind of internal random variables, as close as possible to the standard universe. We apply for the first time to random variables, the general notion of quasi-standard function, as introduced by Robinson and expounded by Luxemburg [15],[17].

If $F_i(y)$ is the cumulative distribution function of $X(i)$, and k is in ${}^*I$ but not in $\dot{I}$, then $F_k(y)$ is a quasi-standard function from ${}^*R$ into the non-standard enlargement of the unit interval.

<u>Definition</u>: We will refer to F_k as the <u>distribution function</u> of $X(k)$.

It follows from the Transfer Principle that F_k exists and has all the properties that are shared by standard distribution functions. It is monotonic non-decreasing and becomes smaller than any positive (standard or non-standard) real number for y sufficiently small (y might need to be assigned actually infinite negative values in order to make F_k less than some chosen positive value.) In a similar sense, F_k approaches 1 as y becomes sufficiently large, and is right-continuous for all y.

Furthermore, F_k shares any property that is shared by all the distribution functions of the standard family $\{X_i\}$. For example, if the X_i are symmetric, so is X_k in the sense that $F_k(y) \equiv 1 - F_k(-y)$.
If the X_i are all normally distributed, with mean $m(i)$ and variance $s^2(i)$, then X_k is normally distributed, with mean $m(k)$ and variance $s^2(k)$. In the simplest case, where $m(i) \equiv 0$, and $s^2(i) \equiv i$, then also $m(k) \equiv 0$, and $s^2(k) \equiv k$; but k, of course, can be infinite, or infinitesimal. The meaning of this is simply that in any formula which is valid for all i, we may substitute the non-standard value k, and obtain a valid formula.

Since $F_i(y)$ is, for standard i and y, the measure of a certain set in the sample space of X_i, there is associated in a natural way with the quasi-standard distribution function F_k a collection of quasi-standard sets in an enlargement of the sample space. This collection is a $\star$-sigma algebra, and is endowed with a $\star$-measure P_k which takes on values in $^*R$ between 0 and 1, possibly including non-zero infinitesimal values. The measure-theoretic study of this $\star$-sigma algebra is related to general questions on the foundations of measure theory -- for example, the reduction of standard measures to atomic measures in non-standard enlargements. These questions are receiving attention from specialists in non-standard analysis (see, e.g., Loeb and Bernstein [2,12-14]. For our purposes, which have more to do with the study of stochastic processes and limit theorems, it seems to be possible to avoid such questions by using the quasi-standard distribution function as our principal tool.

If X and Y are quasi-standard random variables, their joint distribution function is a quasi-standard function of two real variables, defined just as the separate distribution functions are defined. If X_s is the standard family by which X is defined, and Y_r is the standard family by which Y is defined, then the joint distribution for the standard functions is parametrized by the pair (r,s) and letting r and s separately take on non-standard values is the same as letting the ordered pair (r,s) take on a non-standard value as an ordered pair. Thus we can make the definition: X and Y are <u>independent</u> if their joint distribution function is equal to the product of their individual distribution functions; the definition extends immediately to independence of a collection of n quasi-standard random variables, where n may be a finite or infinite natural number.

If w is an infinite natural number (w is in $^*N$ but not in N) and if, for $n \leq w$, X_n are standard or quasi-standard random variables, then $S_w = \sum_{n=1}^{w} X_n$ is a quasi-standard random variable. If the X_n are independent, then the distribution function of S_w is the w-fold convolution product of the distribution functions F_n , $1 \le n \le w$. If the X_n have means m_n , S_w has a well-defined mean $m_w = \sum_{n=1}^{w} m_n$. If they are independent and have variances s_n^2 , then S_w has a variance $s_w^2 = \sum_{n=1}^{w} s_n^2$. m_w and s_w may each be infinite, finite, or infinitesimal. If the X_n are identically distributed, we have simply $m_w = wm$, $s_w^2 = ws^2$. If the X_n are standard, it follows that then m_w and s_w are each either infinite or zero, since m and s are standard, and must be either zero or non-infinitesimal; and the product of an infinite number with a non-infinitesimal number is infinite. On the other hand, if the X_n are quasi-standard, then wm or ws^2 may be infinite, finite or infinitesimal. If the X_n are independent and Gaussian, then S_w is Gaussian. If we define $Z_w = \sum_{n=1}^{w} (X_n - m_n)/s_n w^{1/2}$, then Z is Gaussian with mean zero and variance 1. In that case, Z_w is a quasi-standard random variable whose distribution function is a standard function.

<u>Definition</u>: We will say a quasi-standard random variable X is <u>infinitesimal in probability</u> and write $X \underset{p}{\sim} 0$ if , for all standard y , its distribution function $F(y)$ satisfies $F(y) \sim 0$ for $y < 0$ and $F(y) \sim 1$ for $y > 0$.

<u>Proposition 1.</u> A sequence of standard random variables $\{X_n\}$ converges to zero in probability iff, for every infinite natural number w , X_w is infinitesimal in probability.

<u>Proof</u>: Apply the Convergence Principle to the standard sequence of real numbers defined by $\{r_n\}$ = probability $[|X_n| \ge y]$ where y is an arbitrary standard non-zero real number.

It is essential to recognize that "$X \underset{p}{\sim} 0$" by no means is the same as "$X \sim 0$ almost surely." Indeed, if y is, for example, a standard negative real number, then "$X \sim 0$ a.s." would imply "$X > y$ a.s." which would imply $F(y) = 0$, not merely $F(y) \sim 0$. So, if $F(y)$ is infinitesimal but non-zero for some negative standard y, X is not almost surely ~ 0. In fact, the notion of a random variable which is almost surely infinitesimal does not appear to be a useful one.

It might be thought that if X is infinitesimal in probability, then the mean of X must exist, and be infinitesimal. But this is not so. For instance, let $\{X_{m,a}\}$ denote a family of random variables which take on the value m with probability a, and the value 0 with probability $1-a$. If a is chosen to be infinitesimal, clearly $X_{m,a} \underset{p}{\sim} 0$. Yet the mean, which is ma, may be any finite or infinite $*$-real number, by suitable choice of m. One can also construct quasi-standard random variables which are infinitesimal in probability and for which the mean is undefined, even as an element of ${}^*R$.

<u>Definition</u>: We will say X is <u>finite in probability</u> if, for all negative infinite y, $F(y) \sim 0$, and for all positive infinite y, $F(y) \sim 1$.

<u>Definition</u>: If F is a quasi-standard distribution function, let
$F^{\#}(y) = \lim_{h \searrow 0} {}^{o}(F(y+h))$.

<u>Proposition 2.</u> If F is the quasi-standard distribution function of a quasi-standard random variable X which is finite in probability, then $F^{\#}$ is a standard distribution function.

<u>Proof</u>: We must show that $F^{\#}$ is right-continuous, non-decreasing, $F^{\#}(-\infty) = 0$, and $F^{\#}(\infty) = 1$. Since

$${}^{o}F(z) - {}^{o}F(y) = (F(z)-F(y)) + ({}^{o}F(z)-F(z)) + (F(y)-{}^{o}F(y))$$

and the last two terms on the right are infinitesimal, ${}^{o}F(z)-{}^{o}F(y) \sim F(z)-F(y)$. Since F is a quasi-standard distribution function, $F(z) - F(y)$ is non-negative if $z > y$, and it is either infinitesimal or not infinitesimal. If $F(z) - F(y)$ is positive and not infinitesimal, then ${}^{o}F(z) - {}^{o}F(y)$ is also positive and not infinitesimal. If $F(z) - F(y)$ is infinitesimal, then ${}^{o}F(z) - {}^{o}F(y)$ is also infinitesimal. But then, since ${}^{o}F(z)$ and ${}^{o}F(y)$ are by construction standard real numbers, so is ${}^{o}F(z) - {}^{o}F(y)$ standard, and therefore it must equal zero. In either case, then, we have ${}^{o}F(z) \geq {}^{o}F(y)$.

The assumption that X is finite in probability implies, by the Convergence Principle, that $\lim_{y\to-\infty} {}^{o}F(y) = 0$ and $\lim_{y\to\infty} {}^{o}F(y) = 1$.

Simple examples show that ${}^{o}F$ can fail to be right-continuous, so in order to get a standard distribution function we must go from ${}^{o}F$ to the unique standard right-continuous function which equals ${}^{o}F$ at the points of continuity of ${}^{o}F$; this is $F^{\#}$, and the proof is complete.

It is clear that if X is standard, then it is finite in probability. On the other hand, can one make the stronger statement, that X is finite with probability one? Suppose, for example, that X is a standard Gaussian random variable. Then, for every standard y , the following sentence is true: "Prob $[X > y] \neq 0$" . By the transfer principle, the same sentence must be true if y is chosen to be an infinite positive real number. So there is a positive (infinitesimal) probability that X is infinite, even though X is standard!

The trouble, of course, is that we are talking about two different probability measures, say P and P^* . In the standard universe X takes on only finite values, and of course, the probability that X is finite equals one. In the non-standard universe the same-named random variable can and does take on infinite values, and, referring now to P^* , it becomes false to say X is almost surely finite, or finite with probability one; the best that can be said is that it is finite "in probability" , according to our definition just given.

(It is not hard to see that if X is standard and finite with probability one, then it is bounded; and conversely.)

The Transfer Principle is not violated by this discrepancy, which arises from distinguishing between the standard set of real numbers (all of which are finite) and the non-standard real numbers (some of which are infinite). This distinction is not possible to make within our formal language $\mathcal{L}$, since $^*R$ is by definition the set in the non-standard universe which has the same name as R in the standard universe. The sentences written above, which informally discuss this discrepancy, are not contained in $\mathcal{L}$, and the Transfer Principle does not apply to them.

Given a standard or quasi-standard distribution function $F(y)$, it will be convenient also to define the "absolute distribution function" $G(y)$: for $y \geq 0$, $G(y) = F(-y) + 1 - F(y)$. If F is the distribution function of a random variable X , then $G(y) = \text{prob}\{|X| \geq y\}$. It is to be understood that if we are discussing an indexed family of standard or quasi-standard random variables X_i , then F_i and G_i are indexed correspondingly.

Proposition 3. Given a sequence $\{X_n\}$ of standard random variables, $\{X_n\}$ converges to zero almost surely if, for every standard y and every pair of infinite natural numbers, $w_1 < w_2$,

(2) $$\sum_{n=w_1}^{w_2} G_n(y) \sim 0 \quad .$$

Proof. By the Convergence Principle, (2) implies that the standard infinite series $\sum_{n=1}^{\infty} G_n(y)$ converges, which implies $\{X_n\}$ converges to zero almost surely, by the Borel-Cantelli lemma.

Proposition 4. If X is a quasi-standard random variable with mean m and variance s^2 , then, for any standard or non-standard positive real number y , $G(y)$ satisfies

$$G(y + |m|) \leq s^2/y^2 \; .$$

Proof: Chebychev's inequality is, by the transfer principle, valid for standard or non-standard m , s and y :

$$\text{Prob}(|X-m| \geq y) \leq s^2/y^2 \; .$$

The proposition follows, upon using the definition of G .

Remark. From Proposition 4, it follows that if $m \sim 0$ and $s \sim 0$, then X is infinitesimal in probability. As noted above, the converse is false; $X \underset{p}{\sim} 0$ does not imply $m \sim 0$ or $s \sim 0$. Also, from Proposition 4, it follows that if m and s are finite, then X is finite in probability.

Proposition 5. If F_X and F_Y are distribution functions of standard or quasi-standard random variables X and Y such that $X - Y$ is infinitesimal in probability, then $F_X^{\#} = F_Y^{\#}$.

Proof: Since $\text{Prob}(X-Y \geq h) \sim 0$ for any standard positive h, we have, for any standard real number z,

$$^{o}[\text{Prob}(X \leq z-h)] \leq {}^{o}[\text{Prob}[(Y+(X-Y)) \leq z]] \leq {}^{o}[\text{Prob}(X \leq z+h)]$$

or ${}^{o}F_X(z-h) \leq {}^{o}F_Y(z) \leq {}^{o}F_X(z+h)$. So, at continuity points of F_X, ${}^{o}F_X = {}^{o}F_Y$, which implies $F_X^{\#} = F_Y^{\#}$.

Proposition 6: If $X_1 - Y_1$ and $X_2 - Y_2$ are each infinitesimal in probability, and F_X is the joint distribution function of X_1, X_2, and we define $F_X^{\#} = \lim_{\substack{\epsilon_1 \to 0 \\ \epsilon_2 \to 0}} {}^{o}F_X(X_1+\epsilon_1, X_2+\epsilon_2)$, then $F_X^{\#} = F_Y^{\#}$.

Proof: Similar to Proposition 5.

Stochastic Differentials

If $X(t)$ is a standard or quasi-standard stochastic process for $t \geq 0$, then, for any infinitesimal dt, we make the

Definition: $dx = X(t+dt) - X(t)$.

Clearly dx is, for each standard or non-standard t in $^*R$, a quasi-standard random variable. It is an "infinitesimal increment" obtained by enlarging a family of finite increments $X(t+h) - X(t)$, $0 < h < a$. The Transfer Principle therefore assures us that dx inherits all formal properties shared by the finite increments of $x(t)$ -- if in any true sentence we simply replace the standard positive real number h by the positive infinitesimal dt . dx is, of course, a function of t , and when we wish to make this dependence explicit, we write dx(t) . dx is also a function of a sample point in the sample space of X(t) -- that is, dx is a random variable -- but we never need to indicate explicitly its dependence on the sample point. This is the usual situation in the study of stochastic processes by standard analysis, and it remains true in nonstandard stochastic analysis.

Finally, dx is a function of dt ; if we choose a different infinitesimal dt , we get a different increment dx . In considering this dependence, we will make statements about dx "for all dt" or "for some dt" . In such phrases it is always to be understood that dt is a positive infinitesimal.

Proposition 6a. On any given sample path, a standard stochastic process X(t) is right-continuous iff on that sample path dx is infinitesimal, for all dt .

Proof. Immediate from the convergence principle; see Robinson (17) Theorem 3.4.5.

Definition: We will say a standard process X(t) is right-continuous in probability if, for all standard positive real numbers y ,

$$\lim_{h \searrow 0} \text{Prob}\{|X(t+h)-X(t)| \geq y\} = 0 .$$

Proposition 6b. X(t) is right-continuous in probability iff dx is infinitesimal in probability for all dt.

<u>Proof</u>. Immediate, from the Definition and Proposition 1.

For each t, and fixed dt, dx has a quasi-standard distribution function which we denote by $F_t(y)$.

We will say dx is <u>stationary</u> if $F_t(y)$ is constant with respect to t.

We will say dx is a <u>white noise</u> if $dx(t)$ and $dx(s)$ are independent for all standard s and t, $s \neq t$.

It follows from the Transfer Principle that, if $X(t)$ is a standard process with independent increments, and $0 \leq t_1 < t_2 < t_3 < t_4$ are any members of ${}^*R$, then $X(t_2) - X(t_1)$ is independent of $X(t_4) - X(t_3)$.

Now we are in a position to consider formula (1). If $b(t)$ is a standard Brownian motion, db is, for each infinitesimal dt, a well-defined quasi-standard stationary stochastic process. In fact, by the Transfer Principle, db is, for each t, normally distributed with mean zero and variance dt. db is a white noise, since $b(t)$ has independent increments.

Since db is normally distributed, its value distribution ranges over the whole (extended) real axis; therefore its square, db^2, has a value distribution which ranges over all non-negative (extended) real values. This remark is already enough to show that (1) must be false, since dt is a single fixed infinitesimal. Neither is it true that $db^2 = dt$ almost surely, or even that $\frac{db^2}{dt} - 1$ is almost surely infinitesimal.

In fact, a routine calculation shows that $\frac{db^2}{dt} - 1$ has a standard distribution function, with mean zero and variance 2.

In what sense, then, is formula (1) correct? First of all, it is correct on the average--that is, the expected value of db^2 is dt. Also, it is correct as to order of magnitude-- that is, both $\frac{db^2}{dt}$ and its reciprocal $\frac{dt}{db^2}$ are finite in probability. (See Theorems 4 and 5). Finally, it is correct when summed over t -- this is the formula on quadratic variation, which is proved below (Theorems 7,8 and 9).

Another example of the paradoxes one encounters is suggested by the question, "is db almost surely infinitesimal?" One can answer "no, since the distribution function of db ,

$$(2\pi\ dt)^{-1/2} \int_{-\infty}^{y} \exp(-s^2/2dt)ds$$

is not identically zero for standard $y < 0$, or identically 1 for standard $y > 0$."

On the other hand, one could answer "yes, since $b(t)$ is almost surely continuous, and db is infinitesimal if and only if $b(t)$ is continuous."

The trouble, of course, is in the phrase "almost surely" ; the answer is "no" in the $*$-measure on $^*R$ induced by $F_t(y)$, "Yes" in the standard measure on the standard sample space; and the Transfer Principle is not violated, because the distinction between infinitesimal and non-infinitesimal, like the distinction between finite and infinite mentioned earlier, is an "external" distinction not expressible in the formal language $\mathcal{L}$.

We will avoid the need to keep track of our "almost surelys" by using this phrase henceforward only in reference to standard properties of standard random variables, as in Proposition 3 above.

Our next observation is that dx and $F_t(y)$ lend themselves to neat and succinct characterizations of processes $X(t)$. What's more, it turns out, surprisingly, that if we make the assumption that $X(t)$ is right-continuous, then our results are independent of dt -- one infinitesimal is as good as all infinitesimals.

Let F_t be a standard increasing family of sigma-fields, $X(t)$ a standard stochastic process measurable F_t , and let $E_t(\cdot)$ denote conditional expectation, given F_t . Assume $X(t)$ is right-continuous in the mean; that is, for $s \leq t$,

(3) $$\lim_{h\to 0} E_s(X(t+h)-X(t)) = 0$$

where, as usual, we use h to denote a standard positive real number, so that (3) is a standard condition. Then we have

<u>Theorem 1</u>: Under assumption (3), the following three conditions are equivalent:

(a) $X(t)$ is a martingale

(b) $E_t(dx) = 0$ for all t and all dt

(c) $E_t(dx) = 0$ for all t and some fixed dt .

<u>Proof</u>. (a) $\Longrightarrow$ (b) By definition of a martingale, $E_t(X(t+h)-X(t)) = 0$ for all standard positive t and h . By the Transfer Principle, we have (b) .

(b) $\Longrightarrow$ (c) trivially.

(c) $\Longrightarrow$ (a) Let h and t be arbitrary positive standard real numbers, and dt some positive infinitesimal. Then h/dt exists as an infinite positive *-real number, and there is a uniquely defined infinite natural number w such that $w-1 < h/dt \leq w$. Thus we can write $h = w\,dt-r$ where $0 \leq r < dt$ and so

$$
\begin{aligned}
E_t(X(t+h)-X(t)) &= E_t[\sum_{j=1}^{w} X(t+j\,dt)-X(t+(j-1)dt)-(X(t+h+r)-X(t+h))] \\
&= E_t[\sum_{j=1}^{w} dx(t+(j-1)dt)] - E_t[X(t+h+r)-X(t+h)] \\
&= \sum_{j=1}^{w} E_t(dx(t+j-1)dt)) - E_t[X(t+h+r)-X(t+h)] \\
&= 0 - E_t[X(t+h+r)-X(t+h)] .
\end{aligned}
$$

Now, since r is infinitesimal, the Convergence Principle and the assumed condition (3) imply that the right side of the last equation is infinitesimal. But since the left side is standard, so is the right, and it must equal zero. The proof is complete.

<u>Theorem 2</u>. If $X(t)$ is a standard stochastic process such that, for all dt , dx is infinitesimal in probability -- i.e., $X(t)$ is right-continuous

in probability--then the following three conditions are equivalent:

(a) $X(t)$ has stationary standard (i.e., finite) increments

(b) dx is stationary for all dt

(c) dx is stationary for some dt .

Proof. As in Theorem 1, (a) $\Longrightarrow$ (b) by the Transfer Principle, and (b) $\Longrightarrow$ (c) is trivial. To prove (c) $\Longrightarrow$ (a), we again use $h = w\,dt - r$, and

$$X(t+h) - X(t) = \sum_{j=1}^{w} X(t+jdt) - X(t+(j-1)dt) - X(t+h+r) - X(t+h))$$
$$= \sum_{j=1}^{w} dx(t+(j-1)dt) - (X(t+h+r) - X(t+h)) \ .$$

Let $F(y)$ denote the distribution function of dx : by hypothesis, it is constant with respect to t . Therefore the w-fold convolution power of F is constant with respect to t ; let this function be denoted by $F^{(w)}(y)$. Thus we can write

$$X(t+h) - X(t) = \Delta_1 - \Delta_2 \quad \text{where} \quad \Delta_1 = \sum_{j=1}^{w} dx(t+(j-1)dt)$$

has distribution function $F^{(w)}$, and $\Delta_2 = X(t+h+r) - X(t+h)$ is infinitesimal in probability, because r is infinitesimal and X is right-continuous by assumption. Then, by Proposition 5, the distribution function of $X(t+h) - X(t)$ is equal to $(F^{(w)})^{\#}$, which is independent of t .

Theorem 3. If $X(t)$ is a standard stochastic process such that, for all dt , dx is infinitesimal in probability, then the following three conditions are equivalent:

(a) non-overlapping finite increments of X are independent

(b) for all dt, dx is a white noise

(c) for some dt , dx is a white noise.

Proof. As before, (a) $\Longrightarrow$ (b) is obvious and (b) $\Longrightarrow$ (c) is trivial. To prove (c) $\Longrightarrow$ (a), we take two finite increments, say h_1 and h_2 , and write each as an integer multiple of dt , with infinitesimal error

r_1 or r_2. There results then, as in Theorems 1 and 2, a decomposition

$$X(t_1+h_1)-X(t_1) = X(t_1+w_1 dt) - X(t_1) + X(t_1+h_1) - X(t_1+h_1+r_1)$$
$$= \Delta_1^1 - \Delta_2^1$$

and similarly

$$X(t_2+h_2) - X(t_2) = \Delta_1^2 - \Delta_2^2 .$$

We assume, of course, that the intervals (t_1, t_1+h_1) and (t_2, t_2+h_2) are disjoint. Then the terms Δ_1^1 and Δ_1^2 are sums of non-overlapping increments of width dt, which are independent by hypothesis and so Δ_1^1 and Δ_1^2 are independent. That is, if F_{12} is the joint distribution function of Δ_1^1 and Δ_1^2, and F_1, F_2 respectively are the distribution functions of Δ_1^1, Δ_1^2, then $F_{12} = F_1F_2$. Since Δ_2^1 and Δ_2^2 are infinitesimal in probability, the joint distribution function of $X(t_1+h_1) - X(t_1)$ and of $X(t_2+h_2) - X(t_2)$ is, by Proposition 6, $F_{12}^{\#}$. By Proposition 5, the distribution function of $X(t_j+h_j) - X(t_j)$ is $F_j^{\#}$, $j = 1,2$. But $F_{12} = F_1F_2$ implies ${}^oF_{12} = {}^oF_1\,{}^oF_2$ and therefore $F_{12}^{\#} = F_1^{\#}\,F_2^{\#}$, which completes the proof.

Our theorems 1-3 characterize standard processes in terms of their differentials. The proofs serve as a first illustration of the use of non-standard techniques. We could restate the theorems in standard terminology. For instance, the hypothesis "dx is infinitesimal in probability for (some) (all) dt" is equivalent to "If $y > 0$, $\mathrm{Prob}(|X(t+h_n)-X(t)| > y)$ converges to zero for (some) (all) sequences $h_n \searrow 0$." With such restatements, these theorems are examples of standard theorems (theorems about standard objects) proved by non-standard methods.

Our next considerations relate to more familiar questions in the theory of continuous-time stochastic processes.

<u>Proposition 7</u>. A standard function $f(t)$ is Hölder-continuous of order α iff there exists a standard positive M such that, for all dt, $\left|\frac{df}{dt^\alpha}\right| \leq M$.

<u>Proof:</u> Apply the second sentence of the Convergence Principle to the usual definition of Hölder continuity.

<u>Definition:</u> We will say a standard stochastic process $X(t)$ is "Hölder-continuous of order α in probability from the right" if, for some standard M, $\lim_{h \searrow 0} \text{Prob}\left(\left|\frac{X(t+h)-X(t)}{h^\alpha}\right| \leq M\right) = 1$.

<u>Theorem 4.</u> If $X(t)$ is a standard stochastic process, and if, for some standard nonnegative α and M, $0 \leq \alpha \leq 1$, and some dt

$$m = E(|dx|) \leq Mdt^\alpha$$

and also, for some $\gamma > \alpha$, $s^2 = \text{Var}(dx) \leq Mdt^{2\gamma}$ then X is Holder-continuous of order α in probability.

<u>Proof:</u> By Proposition 4, $G(y)$, the "absolute distribution function" of dx, satisfies $G(y+m) \leq s^2/y^2$ for any y in ${}^*R$. If we choose $y = dt^\alpha$, then $G(dt^\alpha + Mdt^\alpha) \leq Mdt^{2(\gamma-\alpha)}$ or $\text{Prob}(\frac{|dx|}{dt^\alpha} \leq M+1) \leq 1 - Mdt^{2(\gamma-\alpha)}$ which implies, since $\gamma > \alpha$, $\text{Prob}(\frac{|dx|}{dt^\alpha} \leq M+1) \sim 1$, which implies, by the Convergence Principle,

$$\lim_{h \searrow 0} (\text{Prob}(\frac{|X(t+h)-X(t)|}{h^\alpha} \leq M + 1)) = 1,$$

which was to be proved.

<u>Example:</u> If $X(t)$ is a process with stationary, independent increments, mean 0, and variance of $X(1) = 1$, for instance Brownian motion, then $m = 0$, $s = dt$. One can choose $M = 1$, $\gamma = 1/2$, and α any standard positive number such that $\alpha < 1/2$, and conclude that $X(t)$ is continuous (in fact, Holder continuous of order $\alpha < 1/2$) in probability. Of course much more is true for a separable version of the process. We will not explore the question of separability.

<u>Definition.</u> We will say $X(t)$ is, in probability, not Hölder-continuous of order β if, for every finite M,

(4) $$\lim_{h \searrow 0} \text{Prob}\left(\frac{|X(t+h)-x(t)|}{h^{\beta}} \leq M\right) = 0 \ .$$

Theorem 5: Brownian motion is, in probability, not Holder-continuous of order β for any $\beta > {}^1/2$.

Proof: Since the distribution function of $\frac{db}{\sqrt{dt}}$ is a fixed standard continuous function, $\lim_{h \searrow 0} (\text{Prob}(\frac{|db|}{\sqrt{dt}} \leq h) = 0$. Therefore, by the Convergence Principle, if k is any positive infinitesimal, $\text{Prob}(\frac{|db|}{\sqrt{dt}} \leq k) \sim 0$. Choose $k = M(dt)^{\beta - 1/2}$ where M is any standard positive real number. Then $\text{Prob}(\frac{|db|}{dt^{\beta}} \leq M) = \text{Prob}(\frac{|db|}{\sqrt{dt}} \leq k) \sim 0$ which, according to the Convergence Principle, is equivalent to (4) .

Remark: We have shown that $b(t)$ is, in probability, Hölder-continuous of every order $< {}^1/2$ and not Holder continuous of any order $> {}^1/2$. In particular, by setting $\beta = 1$, we see that $b(t)$ is non-differentiable in probability.

(From the local law of the iterated logarithm we know that $b(t)$ is not Hölder continuous of order ${}^1/2$ and therefore that (1) is false.)

Up to this point the only properties of db that we have used are the value of its mean and variance and the fact that the distribution of $db/\sqrt{dt}$ is fixed, standard, and continuous near the origin. The fact that it is a white noise will now permit us to calculate the quadratic variation of $b(t)$.

The intuitive idea is that the quadratic variation Q , being the limit of a sum of squares of increments of $b(t)$, is according to the Convergence Principle, infinitely close to $Q_w = \sum_{k=1}^{w} db_k^2$, which is a sum of infinitely many independent identically distributed quasi-standard random variables. For instance, if all the dt_k are equal, then $dt_k = T/w$, and one expects that some "law of large numbers" should tell us that this sum is infinitely close to $w \cdot E(db^2) = wdt = w\frac{T}{w} = T$. This is in fact correct. What is needed is a "law of large numbers" for quasi-standard sums where the number of terms is a non-standard infinite natural number. For this purpose, we need an inequality of the same kind that is used in proving a standard law of large numbers. By using successively stronger inequalities, we get sharper theorems. The simplest version uses Chebychev's inequality (Proposition 4).

Proposition 8. Suppose dx is a white noise, and $\{t_j\}$, $1 \le j \le w$, is a partition of an interval $[0,T]$: $dt_j = t_j - t_{j-1} > 0$, $t_0 = 0$, $t_w = T$, $dx_j = x(t_j) - x(t_{j-1})$. Let $m_j = E(dx_j)$, $s_j^2 = E(dx_j - m_j)^2$, $r_j^2 = E((dx_j - m_j)^2 - s_j^2)^2$, $Q_w = \sum_{j=1}^{w} (dx_j - m_j)^2 - s_j^2$. Then the "absolute distribution function" $G_w(y) = \mathrm{Prob}(|Q_w| \ge y)$ satisfies $G_w(y) \le \sum_{j=1}^{w} r_j^2/y^2$ for any y in ${}^*R$.

Proof. Each term under the summation sign in the definition of Q_w has mean zero and variance r_j^2. Therefore Q_w has mean zero and variance $\sum_{j=1}^{w} r_j^2$. The conclusion follows by Proposition 4.

In particular, if dt_j is independent of j (equal mesh size) and dx is stationary, we get $G(y) \le wr^2/y^2 = Tr^2/y^2dt$. More generally, we have $G(y) \le \dfrac{T \max r_j^2}{y^2 \min dt_j}$ where the max and min are for $1 \le j \le w$. In this way we get

Theorem 6:

If $\sum_{j=1}^{w} r_j^2 \sim 0$, Q_w is infinitesimal in probability.

Proof: Proposition 8.

If we specialize to Brownian motion, we have r_j^2 as the variance of the square of a Gaussian random variable with mean 0 and variance dt. An elementary computation yields $r_j^2 = 2dt_j^2$, and so

Theorem 7. Let π_j be a sequence of standard partitions of an interval $[0,T]$, $\pi_j = \{t_j^k\}$, $1 \leq k \leq n_j$. Let $\Delta_j^k = t_j^{k+1} - t_j^k$, and let $\Delta b_j^k = b(t_j^{k+1}) - b(t_j^k)$. Then if $\sum_{k=1}^{n_j} (\Delta_j^k)^2 \to 0$, $\sum_{k=1}^{n_j} (\Delta b_j^k)^2$ converges in probability to T as j goes to infinity.

Proof: Theorem 6 and Proposition 1.

Corollary: Let $M_j = \max_k \Delta_j^k$. Then $\sum_k (\Delta b_j^k)^2 \to T$ in probability if $M_j^2 n_j \to 0$.

Example: If $\Delta_j^k = \frac{T}{n_j}$ (equipartitions), then $M_j^2 n_j = \Delta_j^2 \cdot \frac{T}{\Delta_j} = \Delta_j T$, and $\Delta_j \to 0$ implies $\sum_k (\Delta b_j^k)^2 \to T$ in probability.

Our result on convergence in probability corresponds to a weak law of large numbers for a sum of independent infinitesimals. For convergence almost everywhere -- i.e., for a strong law of large numbers -- we use Proposition 3. If, in Proposition 3,

$$\sum_{n=w_1}^{w_2} \sum_{j=1}^{n} r_j^2 \sim 0$$

for all infinite w_1, w_2, we can conclude that $Q_n \to 0$ almost surely.

In the case of Brownian motion, $\sum_{k=1}^{n_j} (dt_j^k)^2 \leq C\, j^{-1-\epsilon}$ for some positive standard ϵ is sufficient. In the simplest case of equipartitions, $dt_j^k = T/n_j$, and for almost sure convergence we need

$$\sum_{j=w_1}^{w_2} 1/n_j \sim 0 . \tag{5}$$

If $n_j \geq c\alpha^j$, $\alpha > 1$, (5) is satisfied. The case $n_j = 2^j$ is the best-known; it was given by Lévy. If $n_j \geq cj^p$, $p > 1$, (5) is satisfied.

Stronger theorems on quadratic variation can be proved by using more information about the distribution functions of the random differentials, and replacing Chebychev's inequality by stronger ones. For example, if X_i are identically distributed and independent, with mean zero, and $E(X_i^r)$ is finite for every standard natural number r, then Hausdorff's inequality (Lamperti (11), p. 42) says that, for any standard positive a and ε, there is a positive δ and c such that

$$\text{(6)} \qquad \text{Prob}\left\{\left|\sum_{i=1}^{n_j} X_i\right| > a n_j^{1/2+\delta}\right\} \leq c n_j^{-1-\varepsilon} \quad .$$

If we set $X_k = \frac{db_k^2}{dt_k} - 1$ then the quasi-standard random variables X_k satisfy the conditions for Hausdorff's inequality, and (6) is satisfied for all n_j in ${}^*N$. If

$$a = y\left((\max dt^k)\, n_w^{1/2+\delta}\right)^{-1},$$

then

$$\text{Prob}\{|\sum_{k=1}^{n_w} (db^k)^2 - dt^k| > y\} \leq \text{Prob}\left\{\left|\sum_{k=1}^{n_w} \frac{(db^k)^2}{dt^k} - 1\right| > \frac{y}{\max dt^k}\right\}$$

$$= \text{Prob}\left\{\left|\sum_{k=1}^{n_w} \frac{(db^k)^2}{dt^k} - 1\right| > a n_w^{1/2+\delta}\right\}$$

$$\leq c\, n_w^{-1-\varepsilon} \quad .$$

If $(\max dt^k \cdot n_w^{1/2+\delta})$ is finite and n_w is infinite, we can then conclude from Proposition 3 that we have almost sure convergence. In the notation of Theorem 7, we have

<u>Theorem 8.</u> If, for some $\delta > 0$, $M_j^2\, n_j^{1+\delta}$ is bounded as $n_j \to \infty$, then $\sum_{k=1}^{n_j} (\Delta\, b_j^k)^2 \to T$ almost surely.

<u>Proof:</u> By the Convergence Principle, boundedness of $M_j^2\, n_j^{1+\delta}$ is equivalent to finiteness of $M_w^2\, n_w^{1+\delta}$ for any infinite natural number w, which, as we have just seen, gives almost sure convergence by Hausdorff's inequality and Proposition 3.

<u>Example:</u> Again reverting to the case of equipartitions $\Delta t_j = M_j = \frac{T}{n_j}$, we conclude that in this case $\Delta t_j \to 0$ is enough to imply $\sum_k (\Delta b_j^k)^2 \to T$ almost surely. There is <u>no</u> requirement on the rate of convergence of Δt_j.

More generally, if $M_j = o(n_j^{-1/2-\epsilon})$ for any standard $\epsilon > 0$, as $n_j \to \infty$, the conclusion follows.

Theorem 8 applies, not only to Brownian motion, but to any standard process, if its increments are stationary and have finite moments of every (finite) order. We can get a different theorem if we use the exponential rate of decay of the distribution function of db^2/dt. We quote an inequality of T. Kurtz:

"For each $j \in N$, let $X_1^j, X_2^j, X_3^j, \cdots$ be independent random variables with mean zero and let $a_1^j, a_2^j, a_3^j, \cdots$ be positive real. Define

$$F(t) = \sup_{j,k} \text{Prob}\{|X_k^j| > t\}$$

and

$$S_m^j = \sum_{k=1}^{m} a_k^j X_k^j .$$

Suppose $F(t) \leq \exp[-\lambda_0(t-R)]$ for all $t > R$. Let $a_j = \sup_k a_k^j$, $A_j = \sum_k (a_k^j)^2 / a_j$. Then there exists a constant $0 < \rho_j < 1$ depending on δ, R, λ_0 and A (increasing as a function of A) such that

$$\text{Prob}\{\sup_m |S_m^j| > \delta\} \leq 2\rho_j^{1/a_j} ."$$

(See Kurtz (8), p. 1882).

For our case, $X_j^k = \frac{(db_j^k)^2}{dt_j^k} - 1$, $a_j^k = dt_j^k$, and $a_j = M_j$ in the notation of Theorem 7 and 8. (Note that Kurtz's X_k^j is our X_j^k.)

We need to have $\sum_j \rho_j^{1/M_j}$ convergent, in order to use our Proposition 3 (i.e., Borel-Cantelli). For this, $\rho_j^{1/M_j} < j^{-1-\epsilon}$ will be sufficient, for any standard $\epsilon > 0$, and this will be the case if

$$M_j < |\log \rho_j|(1+\varepsilon)^{-1}(\log j)^{-1} .$$

We note that

$$A_j = \frac{1}{M_j} \sum_{k=1}^{N_j} (dt_k^j)^2 \leq T ,$$

and that $\rho_j < 1$. This completes the proof of:

Theorem 9. If $\qquad M_j = o(\log j)^{-1}$

then $\lim\limits_{j\to\infty} \sum\limits_k (\Delta b_j^k)^2 = T$ almost surely.

Recently, Dudley [2] proved this result which, as shown by Fernandez de la Vega [4], gives the best possible asymptotic rate for M_j . Our proof uses only the asymptotic behavior at infinity of the distribution of db , whereas Dudley's needs in addition the symmetry of db: $(\text{Prob}\{db \geq y\} \equiv \text{Prob}\{db \leq -y\})$. (See [2] and [3]).

We turn now from quadratic variation to a version of Ito's lemma. We calculate the generator of the expectation semi-group associated with a general diffusion process.

Given two continuous functions $e(x,t)$ and $f(x,t)$ we define

$$dx = e(b,t)db + f(b,t)dt , \tag{7}$$

where $b(t)$ as before is Brownian motion and dt is an infinitesimal increment of fixed length. (7) is a familiar formula in diffusion theory, but in standard treatments it is meaningless until an integration is performed. From our point of view (7) explicitly defines a quasi-standard stochastic process dx , from which a quasi-standard random variable $x(t)$ would be obtained by summation of (infinitely many) increments dx :

$$x(t) = x(0) + \sum_{k=1}^{w} dx_k \quad \text{if} \quad t = \sum_{k=1}^{w} dt_k .$$

If we could show that $x(t)$ is finite, then ${}^{\circ}x(t)$ would be a well-defined standard random variable. Here we pass over the question of finiteness of $x(t)$. Recall that the generator associated with a Markov process $x(t)$ is by definition the operator which operates on a smooth function j according to

$$Aj = \frac{d}{dt} E[j(x)] \Big|_{t=0} .$$

In standard analysis, it is not permissible to interchange the operations $\frac{d}{dt}$ and E (expectation) because $j(x(t))$ is not differentiable as a function of t. However, $E[j(x(t))]$ is differentiable, and therefore (see Robinson [18], Theorem 3.4.9) its derivative is just the standard part of its differential quotient,

$$Aj = {}^{\circ}[\frac{1}{dt} E[j(x(t+dt))-j(x(t))]] .$$

Assuming the standard function j is uniformly of class C^2, we write a two-term Taylor expansion,

$$\begin{aligned}
Aj &= {}^{\circ}[\frac{1}{dt} E[j'dx+ \frac{1}{2} j''dx^2 + o(dx)^2]]\Big|_{t=0} \\
&= {}^{\circ}[\frac{1}{dt} E[j'(edb+fdt) + \frac{1}{2} j''(e^2db^2+2ef\,db\,dt + f^2dt^2) + o(dx)^2]\Big|_{t=0} \\
&= \{ {}^{\circ}[\frac{1}{dt} E(j'e)E(db)] + E(j'f) + \frac{1}{2}\, {}^{\circ}[\frac{1}{dt} E(j''e^2)E(db^2)] \\
&\qquad + {}^{\circ}[E(j''ef)E(db)] + \frac{1}{2}\, {}^{\circ}[dt\, E(f^2)] + {}^{\circ}[E(o(dx)^2)]\} \Big|_{t=0} \\
&= [f \frac{d}{dx} + \frac{1}{2} e^2 \frac{d^2}{dx^2}]j(x) ,
\end{aligned}$$

i.e. $A = f \frac{d}{dx} + \frac{e^2}{2} \frac{d^2}{dx^2}$.

We have used the facts that db is independent of $b(t)$ (and therefore also of e and f), and that $E(db) = 0$, $E(db)^2 = dt$.

Martingale Convergence.

The notion of quasi-standard random variables can also be useful for discrete-time processes. To illustrate, we conclude by presenting a non-standard proof of the martingale convergence theorem, in the L_2-bounded case.

The proof follows that of Garsia [5] but is made somewhat simpler by use of infinite natural numbers. The advantage is that one can construct at the outset the function which must be the limit of the martingale.

Let S_n be a martingale, $ES_n^2 < M < \infty$. For $w_1, w_2 \in {}^*N-N$, we wish to show that ${}^oS_{w_1} = {}^oS_{w_2}$ a.e. Since $E(S_m|\mathcal{F}_n) = S_n$ for all $m \geq n$,

${}^oE(S_{w_1}|\mathcal{F}_n) = S_n = {}^oE(S_{w_2}|\mathcal{F}_n)$. Therefore it suffices to show that $S_n = {}^oE(S_w|\mathcal{F}_n) + {}^oS_w$ for any $w \in {}^*N-N$.

Now we follow, in outline, Garsia's proof.

1. ES_n^2 is nondecreasing: $E(S_n-S_{n-1})^2 = E(E(S_n-S_{n-1})^2|\mathcal{F}_{n-1}) = ES_n^2 - ES_{n-1}^2$.

 Therefore ES_n^2 converges, i.e. $ES_w^2 - ES_n^2 \underset{n\to\infty}{\sim} 0$ for $w \in {}^*N-N$.

2. $E(S_{w_1}-S_{w_2})^2 \sim 0$:

$$E(S_w-S_n)^2 = E(E(S_w-S_n)^2|\mathcal{F}_n))$$

$$= E(E(S_w^2|\mathcal{F}_n) - 2S_nE(S_w|\mathcal{F}_n)+S_n^2) = ES_w^2 - ES_n^2 \underset{n\to\infty}{\sim} 0 .$$

3. There is a sequence $S_{n_k} \ni \sum E(S_w-S_{n_k})^2$ is finite. For each fixed n_k, $(S_{n_k+m} - S_{n_k})^2$ is a submartingale. For each $\epsilon > 0$,

$$P(\max_{w \geq m+n_k \geq n_k} (S_{n_k+m} - S_{n_k})^2 > \epsilon) \leq \frac{1}{\epsilon^2} E(S_w-S_{n_k})^2 .$$

This statement is true for any $r > n_k$ and hence true for $w \in {}^*N-N$.

By the Borel-Cantelli lemma,

(8) $\max_{w \geq m+n_k \geq n_k} (S_{n_k+m}-S_{n_k})^2 > \epsilon$ finitely often (in k) a.s. for all $\epsilon > 0$.

In particular, ${}^o(S_w-S_{n_k})^2 > \epsilon$ finitely often for all $\epsilon > 0$, i.e. $S_w - S_{n_k} \sim 0$ for $n_k \in N^* - N$.

Using (8) again and a triangle argument we see that $(S_w - S_n)^2 > \epsilon$ finitely often for all $\epsilon > 0$, i.e.

$$S_w - S_n \sim 0 \quad \text{for} \quad w \quad \text{and} \quad n \in N^* - N .$$

BIBLIOGRAPHY

[1] Abrahamse, Allan F., Some applications of nonstandard analysis to the theory of stochastic processes, Preprint No. 35, Dept of Mathematics, University of Southern California, February 1973.

[2] Bernstein, A. and Loeb, P.A., A nonstandard integration theory for unbounded functions, Victoria Symposium on nonstandard analysis, edited, A.E. Hurd and P.A. Loeb, Springer-Verlag Lecture Notes in Math., No. 369, 40-49.

[3] Dudley, R.M., Sample functions of the Gaussian process, Annals of Prob., 1(1973), 66-103.

[4] Fernández de la Vega, W., On almost sure convergence of quadratic Brownian variation, Ann. of Prob., (1974), pp. 551-552.

[5] Garsia, A.M., Topics in Almost Everywhere Convergence, Chicago, 1970.

[6] Hanson, D.L. and Wright, F.T., A bound on tail probabilities for quadratic forms in independent random variables, Annals of Math. Stat., 42 (1971), 1079-1083.

[7] Hersh, Reuben, Brownian motion and nonstandard analysis, Tech. Report 277, Dept. of Mathematics, UNM, May 1973.

[8] Ito, K., Stochastic differentials of continuous local quasi-martingales, stability of stochastic dynamical systems, Lecture Notes in Math., Springer, 294 (1972), 1-7.

[9] Ito, K., Stochastic differentials, to appear in Appl. Math. and Optimization.

[10] Kurtz, Thomas G., Inequalities for the law of large numbers, Annals of Math. Stat., 6 (1972), 1874-1883.

[11] Lamperti, John, Probability, W.A. Benjamin, Inc., New York, Amsterdam, 1966.

[12] Loeb, Peter A., Conversion from nonstandard to standard measure spaces and applications in probability theory, preprint.

[13] Loeb, P.A., A nonstandard representation of measurable spaces, L_{∞} and L^*_{∞}, Contributions to nonstandard analysis, edited by W.A.J. Luxemburg and A. Robinson, North-Holland, 1972, 65-80.

[14] Loeb, P.A., A nonstandard representation of Borel Measures and σ-finite measures, Victoria Symposium on nonstandard analysis, edited by A.E. Hurd and P.A. Loeb, Springer-Verlag Lecture Notes in Mathematics, No. 369, 144-152.

[15] Luxemburg, W.A.J., Nonstandard analysis. Lectures on A. Robinson's theory of infinitesimals and infinitely large numbers, Pasadena, 1962.

[16] Muller, D.W. Nonstandard proofs of invariance principles in probability theory, Appl. of Model Theory to Algebra, Analysis and Probability, Holt, Rinehart and Winston, 1969.

[17] Robinson, A., Introduction to model theory and to the metamathematics of algebra, Studies in Logic and the Foundations of Mathematics, Amsterdam, 1963.

[18] Robinson, A., Non-Standard Analysis, North-Holland, Amsterdam, 1970.

A RANDOM PRODUCT OF MARKOVIAN SEMI-GROUPS OF OPERATORS

Frank J.S. Wang

Abstract

Let $Y(t)$ be a continuous time pure jump process with state space $S = \{1,2,\ldots,n\}$ and let $\xi_0, \xi_1, \ldots,$ be the succession of states visited by $Y(t)$, $\Delta_0, \Delta_1, \ldots$ the sojourn times in each state, $N(t)$ the number of transitions before t and $\Delta_t = t - \sum_{k=0}^{N(t)-1} \Delta_k$. For each $k \in S$, let $T_k(t)$ be an operator semigroup on a Banach space L with infinitesimal generator $\mathcal{A}_k$. Define $T_\lambda(t,\omega) = T_{\xi_0}(\frac{1}{\lambda}\Delta_0)T_{\xi_1}(\frac{1}{\lambda}\Delta_1)\cdot \ldots \cdot T_{\xi_{N(\lambda t)}}(\frac{1}{\lambda}\Delta_{\lambda t})$. It is known (Kurtz) that if

$$\lim_{t \to \infty} \frac{1}{t}\int_0^t \delta_k(Y(s))ds = \mu_k$$

exist for all $k = 1,2,\ldots,n$ and $\sum_{i=1}^{n}\mu_i = 1$, then under appropriate conditions the closure of $\mathcal{A} = \sum_{i=1}^{n}\mu_i\mathcal{A}_i$ is the infinitesimal operator for a strongly continuous semigroup $T(t)$ defined on L and $T_\lambda(t,\omega)$ converges almost surely to $T(t)$ as $\lambda \to \infty$. The existence and identification of this limit is of interest even when the closure of $\mathcal{A}$ is not a generator. A probabilistic version of this problem is given in the case of Markovian transition semi-group when the corresponding processes have identical hitting distributions. Sufficient conditions for the existence of limit are given. With $S = \{1,2\}$ and $Y(t) = \begin{cases} 1, & 2n \le t < 2n+1 \\ 2, & 2n+1 \le t < 2n+2 \end{cases}$, $n = 0,1,2,\ldots,$ the result is obtained by Loren Pitt.

1. Introduction

Let $Y(t)$ be a stochastic process with values in a finite state space $S = \{1, 2, \ldots, n\}$. We assume that the sample space $\Omega = D[0, \infty)$, the space of right continuous functions with left hand limits taking values in S and $Y(t, \omega) = \omega(t)$.

Under this assumption it makes sense to talk about $\xi_0, \xi_1, \ldots$, the sequence of states visited, and $\Delta_0, \Delta_1, \ldots$, the sojourn times in these states. We define $N(t)$ to be the number of transitions before time t and

$$\Delta_t = t - \sum_{k=0}^{N(t)-1} \Delta_k.$$

Where Σ_0^{-1} is considered to be 0.

For each $i \in S$, let $T_i(t)$ be a semigroup of linear operators on a Banach space L with infinitesimal operator $\mathcal{A}_i$ satisfying $\|T_i(t)\| \leq e^{\alpha t}$ for some fixed α. Use $B(L)$ to denote the space of bounded linear operators on L. Griego and Hersh [2] defined the random evolution $T_\lambda(t, \omega) = [0, \infty) \times \Omega \to B(L)$ governed by $Y(\lambda t)$ as

(1-1) $$T_\lambda(t, \omega) = T_{\xi_0}(\tfrac{1}{\lambda}\Delta_0)\cdot \ldots \cdot T_{\xi_{N(\lambda t)}}(\tfrac{1}{\lambda}\Delta_{\lambda t}),$$

which is called a "random Trotter product" by Kurtz. We are interested in the behavior of $T_\lambda(t, \omega)$ as λ tends to infinity, that is, in what happens if the mode of development of the random evolution changes at a very rapid rate.

Let $\delta_1, \delta_2, \ldots, \delta_n$ denote Kronecher's function at 1, 2, ..., n, respectively, i.e., $\delta_i(j) = 0$ if $i \neq j$ and 1 otherwise. We assume that there exists a positive measure μ on S such that $\sum_{i=1}^{n}\mu(i) = 1$ and

(1-2) $$\lim_{t \to \infty} \frac{1}{t}\int_0^t \delta_i(Y(s))ds = \mu(i) = \mu_i$$

exists almost surely for all $i = 1, \ldots, n$. (W.l.o.g. we also assume that $\mu_i > 0$ for all $i \in S$.) Since S is finite, the assumption above is equivalent to Kurtz's assumption (2-2) in [3]. Define $\mathcal{a}f = \sum_{i=1}^{n} \mu_i \mathcal{a}f$. Kurtz proved that, if the range of $\lambda - \mathcal{a}$ is dense in L for some $\lambda > \sigma$, then the closure of $\mathcal{a}$ is the infinitesimal operator for a strongly continuous semigroup T(t) defined on L and

$$(1\text{-}3) \qquad P\left\{ \lim_{\lambda \to \infty} T_\lambda(t, \omega) = T(t)f \right\} = 1$$

for every $f \in L$.

Pitt [4] considered the case $S = \{1, 2\}$,

$$(1\text{-}4) \qquad Y(t) = \begin{cases} 1 & 2n \leq t < 2n + 1 \\ 2 & 2n + 1 \leq t < 2n + 2 \end{cases}$$

$n = 0, 1, 2, \ldots$, and $T_1(t)$, $T_2(t)$ are two strongly continuous contraction semigroups. He gave an example of uniform translations in opposite directions which shows that the Trotter product

$$(1\text{-}5) \qquad (T_1\ T_2)(t) = \lim_{h \to 0} (T_1(h)\ T_2(h))^{[\frac{t}{h}]}$$

exists and equals the identity operator I, while the closure of $\mathcal{a}_1 + \mathcal{a}_2$ is not a generator. Pitt treats this problem (the existence of Trotter product) probabilistically in the case of Markovian transition semigroups with identical hitting distribution. He proved that the limit (1-5) exists provided that the time change is not too singular.

In this paper, we give a probabilistic analysis of the convergence of $T_\lambda(t, \omega)$ for a class of Markovian semigroups. We assume that $\{T_i(t) : i = 1, 2, \ldots n\}$ are Hunt semigroups on the same state space E and that the $T_i(t)$ process $X_i(t)$ is obtainable from the $T_1(t)$ process $X_1(t)$ by a random time change corresponding to the functional A_i (see Sec. 2 for definition). We will find conditions (on A_i) under which $T_\lambda(t, \omega)$ will converge bounded point-wisely almost surely as λ tends to infinity for all bounded continuous function f on E. (Theorems 3-16 and 3-19.)

2. Definitions and Notations

Let E denote a locally compact metric space, and $\mathcal{B}$ denote its Borel σ-algebra. Let $P(t, x, \Gamma)(t \geq 0, x \in E, \Gamma \in \mathcal{B})$ be a transition function on the measurable state space $(E, \mathcal{B})$ which is conservative, i.e., $P(t, x, E) = 1$ for all $x \in E$ and $t \geq 0$. Let $\overline{\mathcal{B}}$ denote the completion of the σ-algebra $\mathcal{B}$ with respect to the system of all finite measures μ. Every measure $P(t, x, \cdot)$ can be extended uniquely to the σ-algebra $\overline{\mathcal{B}}$. It is easy to verify that such an extension gives a transition function on the state space $(E, \overline{\mathcal{B}})$. Thus we may assume that $\mathcal{B} = \overline{\mathcal{B}}$.

For convenience, we take our basic sample space to be the space Ω^* of right continuous functions $\omega^*:[0, \infty) \to E$ and consider only those non-terminal processes. To denote the process defined on Ω^* we write $X(t)(\omega^*) = X(t, \omega^*) = \omega^*(t)$. The basic σ-algebra $\mathcal{J}$ and $\mathcal{J}_t$ $(t \geq 0)$ on Ω^* are those generated by the sets $\{X(s) \in B\}$ for $B \in \mathcal{B}$ and $s \geq 0$ (respectively $t \geq s \geq 0$). The shift operator θ_t is given on Ω^* by

$$X(s)(\theta_t \omega^*) = X(s + t)(\omega^*).$$

Suppose $P(t, x, \Gamma)$ admit a realization $(X(t))$, that is, we assume for each initial distribution μ on E, there exists a probability measure P_μ on $(\Omega^*, \mathcal{J})$ satisfying the following three properties:

(2-1) $P_\mu\{X(0) \in B\} = \mu(B)$ for each $B \in \mathcal{B}$;

(2-2)   For each $\mathcal{B}$-measurable $f \geq 0$ and $0 \leq s \leq t$

$$E_\mu[f(X(t))|\mathcal{J}_s] = \int_E f(x) \cdot P(t - s, X(s), dx) \text{ a.s.;}$$

(2-3)   For each A in $\mathcal{J}$, $P_x(A)$ is a $\mathcal{B}$-measurable function on x and satisfies

$$P_\mu(A) = \int_E P_x(A)\mu(dx),$$

where P_x is the measure corresponding to the initial point distribution at $x \in E$.

Complete the σ-algebra $\mathcal{F}$ with respect to the measure P_μ. Taking the intersection over all μ, we obtain a new σ-algebra $\eta \supseteq \mathcal{F}$. Let $\mathcal{F}_t^\mu$ be the σ-algebra consisting of all the elements of η which differ from an element of $\mathcal{F}_t$ only by a P_μ null set. Put $\eta_t = \bigcap_\mu \mathcal{F}_t^\mu$. Then the process $(X(t))$ also has the Markov property (2-2) with respect to the σ-algebra (η_t). A stopping time of (X_t) is a η-measurable function $\tau : \Omega^* \to [0, \infty)$ such that $\{\tau < t\} \in \eta_t$ for each $t \geq 0$.

We assume that $(X(t))$ is a Hunt process, i.e., it is a strong Markov process such that

(2-4) for each increasing sequence $\tau_n \uparrow \tau$ of stopping times, we have $X(\tau_n) \to X(\tau)$ a.s. on $\{\tau < \infty\}$ as $n \to \infty$.

We further assume that $X(t)$ is independent of $Y(t)$. For each $i \in S$, let $A_i = \{A_i(t) : t \geq 0\}$ be a continuous additive functional (caf) of the Hunt process $(X(t))$, i.e., A_i satisfies the following three conditions:

(i) for every t, $A_i(t) \in \mathcal{F}_t$;

(ii) almost surely the mapping $t \to A_i(t)$ is nondecreasing right continuous and satisfies $A_0 = 0$;

(iii) for each t and s, $A_i(s + t) = A_i(t) + A_i(s) \cdot \theta_t$ almost surely.

Define the functional inverse $\alpha_i(t)$ of A_i as follows: $\alpha_i(t, \omega^*) = \inf\{s : A_i(s)\omega^*) > t\}$ if such s exists or $+\infty$ otherwise. It is easy to see that, for each t, $\alpha_i(t)$ is a stopping time and it is right continuous as a function of t. Define $X_i(t) = X(\alpha_i(t))$ as the process obtained from $X(t)$ by a random time change corresponding to the functional A_i and $T_i(t)$ as the corresponding semigroup defined on $B(E, \mathcal{B})$, the space of bounded measurable functions. The processes $\{X_i(t)\}_{i=1}^n$ are also Hunt processes (see [1]), and for each $f \in B(E, \mathcal{B})$

$$(2\text{-}5) \qquad T_i(t)f(x) = E_x[f(X_i(t))] = E_x[f(X(\alpha_i(t)))].$$

For each $\omega \in \Omega$ and $\lambda > 0$, we introduce the stopping time $\{\gamma_{\lambda,k}\}_{k=0}^{\infty}$ and $\nu(\lambda, t)$ by

$$(2\text{-}6) \qquad \gamma_{\lambda,0} = \alpha_{\xi_0}\left(\frac{\Delta_0}{\lambda}\right)$$

$$\gamma_{\lambda,1} = \gamma_{\lambda,0} + \alpha_{\xi_1}\left(\frac{\Delta_1}{\lambda}\right)\theta(\gamma_{\lambda,0})$$

$$\vdots$$

$$\gamma_{\lambda,k} = \gamma_{\lambda,k-1} + \alpha_{\xi_k}\left(\frac{\Delta_k}{\lambda}\right)\theta(\gamma_{\lambda,k-1})$$

and

$$\nu(\lambda, t) = \gamma_{\lambda,N[\lambda t]-1} + \alpha_{\xi_{N[\lambda t]}}\left(\frac{\Delta_{\lambda t}}{\lambda}\right)\theta(\gamma_{\lambda,N[\lambda t]-1}).$$

A repeated application of the strong Markov property yields

$$(2\text{-}7) \qquad T_{\xi_0}\left(\frac{\Delta_0}{\lambda}\right) \cdots T_{\xi_{N[\lambda t]}}\left(\frac{\Delta_{\lambda t}}{\lambda}\right)f(x) = E_x[f(X(\nu(\lambda, t)))].$$

Equation (2-7) enables us to make arguments involving the stopping times $\{\nu(\lambda, t)\}$ rather than the semigroup $T_i(t)$. We will find conditions which guarantee the convergence of $\nu(\lambda, t)$ as $\lambda \to \infty$, and this gives corresponding theorems about the random product $T_\lambda(t, \omega)$.

3. Main Theorems

To simplify the notation, we will assume $A_1(t) \equiv I(t) \equiv t$, i.e., $X_1(t) = X(\alpha_1(t)) = X(t)$.

Set $B(t) = \sum_{i=1}^{n}\mu_i A_i(t)$. For each sample path $\omega^* \in \Omega$, the Radon-Nikodyn theorem guarantees the existence of non-negative functions $a_1(t), \ldots, a_n(t)$ with $\sum_{i=1}^{n}\mu_i a_i \equiv 1$ and such that

(3-1) $$A_i(t) = \int_0^t a_i(s)dB(s).$$

Note that since $A_1(t) \equiv t$ by assumption, $dB(s) = \frac{ds}{a_1(s)}$ on the set $\{s = a_1(s) \neq 0\}$. Put $\prod_{i=1}^{n} a_i(s) = a(s)$, $\frac{a(s)}{a_i(s)} = 1$ if $a_1(s) = 0$ and $\sum_{i=1}^{n}\mu_i \cdot \frac{a(s)}{a_i(s)} = b(s)$. The functional

(3-2) $$A(t) = \int_0^t \frac{a(s)}{b(s)} dB(s)$$

is easily seen to be a caf of the process $(X(t))$.

Lemma 3-3: For each sample path ω^*, $A(t)(\omega^*) \leq B(t)(\omega^*)$.

Proof: Since $\sum_{i=1}^{n}\mu_i a_i = 1$ and $0 \leq a_i \leq 1$, Jensen's inequality implies $(\sum_{i=1}^{n}\mu_i \frac{1}{a_i}) \geq 1$. Thus

$$A(t) = \int_0^t (\sum_{i=1}^{n}\mu_i(\frac{1}{a_i}))^{-1} dB(s) \leq B(t).$$

In the case of non-singular diffusion processes on $E = R^1$, the definition of $A(t)$ becomes more explicit. Let $X_0(t)$ denote the one-dimensional Brownian motion on R^1 and let $X_1, \ldots, X_n$ be diffusions on R^1 with zero drift and generator

$$\frac{1}{2}\sigma_1^2(x)\frac{d^2}{dx^2}, \ \ldots, \ \frac{1}{2}\sigma_n^2\frac{d^2}{dx^2},$$

respectively. The speed measure of X_i is given on $\mathcal{B}_1$ by

$$m_i(dx) = \frac{1}{\sigma_i^2(x)}dx,$$

and

$$A_i(t) = \int_0^t \frac{1}{\sigma_i^2(X_0(\xi))}d\xi.$$

In this case it is easy to verify that

$$A(t) = \int_0^t (\sum_{i=1}^{n}\mu_i \sigma_i^2(X_0(\xi)))^{-1} d\xi.$$

For each $i \in S$ and each positive integer K, let $\tau_i(k)$ denote the total number of transitions that $Y(t)$ has taken to accomplish the kth visit to state i. To simplify the notation, we will drop the subscript when $i = 1$ and write τ instead of τ_1. For each state i, denote the total sojourn time in state i before the kth transition by $S_i(k)$, i.e.,

$$(3\text{-}4) \qquad S_i(k) = \Sigma \Delta_{\tau_i(j)},$$

where the summation is taken over the set $\{j = \tau_i(j) < k\}$. Define

$$S(k) = \sum_{i=0}^{k-1} \Delta_i$$

as the time that the kth transition occurs. Note that $S(\tau(k))$ is the time that the process $Y(t)$ makes its kth visit to state 1. Define a family $C_\lambda(t)$ of non-additive functionals on $\{X(t)\}$ as follows:

$$(3\text{-}5) \qquad C_\lambda(t) = \frac{1}{\lambda} S(k) + A_{\xi_k}(t) - A_{\xi_k}(\gamma_{\lambda,k-1}) \text{ if}$$

$$\gamma_{\lambda,k-1} \leq t < \gamma_{\lambda,k}.$$

Let $M \subseteq \Omega$ be the set of all $\omega \in \Omega$ satisfying condition (1-2). The following is basic.

<u>Theorem 3-6</u>: There exists a set $\Omega' \subset \Omega^*$ with $P_x(\Omega') = 1$ for each $x \in E$ and such that on Ω'

$$\lim_{\lambda \to \infty} C_\lambda(t) = A(t)$$

for each $t \geq 0$ and each $\omega \in M$.

This theorem is proven in Section 4. In the remainder of this section we discuss the implications when applied to the random product $T_\lambda(t, \omega)$. All the corollaries are given (in the case where $Y(t)$ is given by (1-4) by Pitt in [4]. We will only give proofs to those that are either not given by Pitt or are not direct consequences of his results.

Corollary 3-7: Let $\gamma(t)$ be the functional inverse of A. Then for each $x \in E$, $t \geq 0$ and $\omega \in M$.

$$\lim_{\lambda \to \infty} \nu(\lambda, t) = \gamma(t) \qquad \text{a.s. } (P_x)$$

on the set of ω^* on which $\gamma(t)$ is continuous at t.

Proof: Let $V_1(t)$ denote the total number of visits of $Y(s)$ to state 1 before time t. If $\tau(V_1(\lambda t)) = N[\lambda t]$, then $\xi_{N[\lambda t]} = 1$ and $\gamma_{\lambda, N[\lambda t]-1} \leq \nu(\lambda, t) < \gamma_{\lambda, N[\lambda t]}$. Therefore it follows from (3-5) and the assumption $A_1(t) \equiv t$ that

$$C_\lambda(\nu(\lambda, t)) = t \quad \text{for all } \lambda, t \text{ such that } \tau[V_1(\lambda t)] = N[\lambda t]. \tag{3-8}$$

If $\tau(V\ (\lambda t)) < N[\lambda t]$, put $\underline{h}(\lambda, t) = S(\tau(V\ (\lambda t)) + 1$ and $\overline{h}(\lambda, t) = S(\tau(V\ (\lambda t)+1))$. Then

$$Y(s) \neq 1 \quad \text{for every s such that } \underline{h}(\lambda, t) \leq s < \overline{h}(\lambda, t) \tag{3-9}$$

and

$$\frac{1}{\lambda}\underline{h}(\lambda, t) = C_\lambda(\gamma_{\lambda, \tau(V\ (\lambda t))}) \leq C_\lambda(\nu(\lambda, t)) \leq C_\lambda(\gamma_{\lambda, \tau(V\ (\lambda t)+1)-1}) = \frac{1}{\lambda}\overline{h}(\lambda, t). \tag{3-10}$$

Now (3-9) implies

$$\frac{1}{\overline{h}(\lambda, t)} \int_0^{\overline{h}(\lambda, t)} \delta_1(Y(s, \omega))ds = \frac{\lambda t}{\overline{h}(\lambda, t)} \cdot \frac{1}{\lambda t} \int_0^{\lambda t} \delta_1(Y(s, \omega))ds. \tag{3-11}$$

Thus

$$\frac{\lambda t}{\overline{h}(\lambda, t)} \to 1 \text{ as } \lambda \to \infty \text{ for all } \omega \in M. \tag{3-12}$$

Similarly we have

(3-13) $\frac{\lambda t}{\underline{h}(\lambda, t)} \rightarrow 1$ as $\lambda \rightarrow \infty$ for all $\omega \in M$.

It is easy to see that (3-8), (3-10), (3-12) and (3-13) imply

(3-14) $\lim_{\lambda \to \infty} C_\lambda(\nu(\lambda, t)) = t$

as $\lambda \to \infty$. The corollary follows from (3-14) and theorem 3-6.

Corollary 3-15: If $\gamma(t)$ is a.s. (P_x) continuous at t, then

$$X(\nu(\lambda, t)) \longrightarrow X(\gamma(t)) \quad \text{a.s. } (P_x)$$

on $\{\nu(t) < \infty\} \times M$.

Theorem 3-16: Suppose the process Y(t) satisfies condition (1.2), a.s., that is, $P(M) = 1$ and that for each fixed $t > 0$ and $x \in E$, γ is continuous at t, a.s. (P_x) (i.e., $\gamma(t)$ has no fixed discontinuities). If also $A(\infty) = \infty$ a.e., then $\gamma(t) < \infty$ a.s. and

$$P\{\omega: \lim_{\lambda \to \infty} X(\nu(\lambda, t)) = X(\gamma(t)) \text{ a.s. for each } t \geq 0\} = 1.$$

Moreover,

(3-17) $P\{\omega: \lim_{\lambda \to \infty} T_\lambda(t, \omega) f(x) = E_x[f(X(\gamma(t)))] \ \forall\ x \in E$ and $t \geq 0$ and every bounded continuous f on E $\} = 1$.

Remark: Put $T_t f(x) = E_x[f(X(\gamma(t)))]$ and $\mathcal{A} f = \lim_{t \to 0} \frac{T_t f - f}{t}$ as the semigroup of operators and generator corresponding to the process $X(\gamma(t))$ respectively, then (3-17) is equivalent to

(3-18) $P\{\omega: \lim_{\lambda \to \infty} T_\lambda(t, \omega) f = T(t) f\} = 1$

for every f bounded and continuous on E if $\gamma(t) < \infty$ a.s., where the convergence in (3-18) is considered to be bounded point-wise, even when the closure of $\sum_{i=1} \mu_i \mathcal{A}_i$ does not generate a semigroup.

Consider the case of non-singular diffusions on R^1. Let

$m_1, m_2, \ldots, m_n$ be n Borel measures on R^1 such that $0 < m(J) < \infty$ for all finite open intervals J. Let X_0 denote the standard Brownian motion on R^1 with the local time at x denoted by $\mathfrak{t}(t,x)$. Let $X_k(t)$ be the diffusions corresponding to the generator $D_{m_k} D_x$. Put

$$d\mu = \frac{dm_1 \ldots dm_n}{\sum_{i=1}^{n} \mu_i \, dm_1 \ldots \widehat{dm_i} \ldots dm_n}$$

where "^" denotes the missing term in the product. Denote the functional inverse of the caf

$$\int_{R^1} \mathfrak{t}(t,x) d\mu(x)$$

of the Brownian motion $X_0(t)$ by $\alpha(t)$.

Theorem 3-19: Let $\{X_k(t)\}$ be the diffusion corresponding to the generators $\{D_{m_k} D_x\}$ defined as above. Suppose the process $Y(t)$ satisfies condition (1-2) a.s., $d\mu \neq 0$ and T_λ is defined as in (1-1), then

$$P\{\omega \mid \lim_{\lambda \to \infty} T_\lambda(t,\omega) f(x) = E_x[f(X_0(\alpha(t)))] \text{ for every } x \in E \text{ and } t \geq 0\} = 1$$

for every bounded continuous f on E.

4. The Proofs

We will continue to assume that $A_1(t) \equiv t$. Define a new sequence of stopping time $\{\nu_{\lambda,k}^*\}$ by

(4-1) $$\nu_{\lambda,k}^* = \nu_{\lambda,\tau(k)} \quad \text{for all } k = 1,2,\ldots$$

Let $\Omega' \subseteq \Omega^*$ be the set of all ω^* such that $A_i(t)(\omega^*)$ is continuous and non-decreasing. We will prove that

(4-2) $$\lim_{\lambda \to \infty} C_\lambda(t) = A(t)$$

for all $t \geq 0$, $\omega \in M$ and $\omega^* \in \Omega'$. From now on we will assume that $\omega \in M$ and $\omega^* \in \Omega'$ are fixed. For each $t \geq 0$, introduce the sets

$$(4\text{-}3) \qquad \begin{aligned} G_\lambda &= \{s \in [0,t] \mid s \in [v^*_{\lambda,k}, v^*_{\lambda,k} + \tfrac{\Lambda_{\tau(k)}}{\lambda}) \text{ for some } k\} \\ G^i_\lambda &= \{s \in [0,t] \mid s \in [v_{\lambda,j}, v_{\lambda,j+1}) \text{ where } v_{\lambda,j+1} = \\ &\qquad v_{\lambda,j} + \alpha_i(\tfrac{\Delta_j}{\lambda})\cdot\theta(v_{\lambda,j}) \text{ for some } j\} \end{aligned}$$

$i = 2,\ldots,n$. Note that from the definition above

$$(4\text{-}4) \qquad G_\lambda = [0,t] - \bigcup_{i=2}^{n} G^i_\lambda.$$

The proof of theorem 3-6 strongly depends on the next lemma.

<u>Lemma 4-5</u>: Let $\omega \in M$ and $\omega^* \in \Omega'$ be fixed. Given any interval $I = [a,b]$, let θ_λ be the number of disjoint intervals in $G_\lambda \cap I$. Denote the first interval in $G_\lambda \cap I$ by $[v^*_{\lambda,m}, v^*_{\lambda,m} + \frac{\Lambda_{\tau(m)}}{\lambda})$ and the last interval in $G_\lambda \cap I$ by $[v^*_{\lambda,m+\theta_\lambda-1}, v^*_{\lambda,m+\theta_\lambda-1} + \frac{\Lambda_{\tau(m+\theta_\lambda-1)}}{\lambda})$.

Let $S_i(k)$ denote the total sojourn time in state i before k^{th} transition as before. Then

$$(4\text{-}6) \qquad \frac{1}{\lambda}(S_i(\tau(m+\theta_\lambda-1)) - S_i(\tau(m)) = \frac{\mu_i}{\lambda}(S(\tau(m+\theta_\lambda-1)) - S(\tau(m))) + O(\tfrac{1}{\lambda})$$

for all $i \in S$ where $O(\frac{1}{\lambda})$ tends to zero as $\lambda \to \infty$ uniformly over all $I \subseteq [0, t]$.

<u>Remark 4-7</u>: (i) In (4-6), m and θ_λ are integer-valued functions of $\omega^* \in \Omega'$ and $S_i(\tau(k))$ is a positive real valued function on $\omega \in \Omega$.

(ii) It follows from this lemma and the proof of Corollary 3-7 that

$$C_\lambda(G^i_\lambda \cap I) = \mu_i C_\lambda(I) + O(\tfrac{1}{\lambda})$$

and

$$C_\lambda(G_\lambda^i \cap I) = \frac{\mu_i}{\mu_1} C_\lambda(G_\lambda \cap I) + O(\tfrac{1}{\lambda})$$

when $O(\frac{1}{\lambda})$ tends to zero uniformly over all $I \subseteq [0,t]$.

Proof: Let $i \in S$ be fixed. Since

$$\frac{1}{\epsilon}\int_t^{t+\epsilon} \delta_i(Y(\lambda s,\omega))ds = \frac{1}{\lambda\epsilon}\int_{\lambda t}^{\lambda(t+\epsilon)} \delta_i(Y(s,\omega))ds$$

is uniformly continuous in t, (3-1) implies

$$P\left\{\lim_{\lambda\to\infty} \sup_{\eta>\epsilon} \sup_{t\leq T} \left|\frac{1}{\eta}\int_t^{t+\eta} \delta_i(Y(\lambda s,\omega)ds - \mu_i\right| = 0\right\} = 1$$

for every $\epsilon > 0$ and $T > 0$. Let $\epsilon_n \to 0$, $T_n \to \infty$ and $\delta_n \to 0$. Then there exists a sequence λ_n such that

$$\sup_{i\leq n} P\left\{\sup_{\lambda\geq\lambda_n} \sup_{\eta\geq\epsilon_n} \sup_{t\leq T_n} \left|\frac{1}{\epsilon_n}\int_t^{t+\epsilon_n} \delta_i(Y(\lambda s,\omega))ds - \mu_i\right| > \delta_n\right\} \leq \delta_n.$$

Setting $\epsilon(\lambda) = \epsilon_n$ for $\lambda_n \leq \lambda < \lambda_{n+1}$, then $\lim_{\lambda\to\infty} \epsilon(\lambda) = 0$ and

$$\text{(4-8)}\qquad P\left\{\lim_{\lambda\to\infty} \sup_{\epsilon\geq\epsilon(\lambda)} \sup_{t\leq T} \left|\frac{1}{\epsilon}\int_t^{t+\epsilon} \delta_i(Y(\lambda s,\omega))ds - \mu_i\right| = 0\right\} = 1$$

for every $T > 0$.

Since $\frac{1}{t}\int_0^t \delta_1(Y(s,\omega))ds \to \mu_1$ a.s., for each $\omega \in M$, there exists an $T'(\omega) > 0$ such that

$$\text{(4-9)}\qquad \frac{1}{t}\int_0^t \delta_1(Y(s,\omega))ds \geq \frac{\mu_1}{2}$$

for every $t > T'$. Put $T = \frac{2b}{\mu_1} + T'$. Since $S_1(\tau(m)) \leq b\lambda$, (4-9) implies

$$\text{(4-10)}\qquad S(\tau(m)) < \lambda T$$

for all $\lambda > 0$ and all $\omega^* \in \Omega$ (if $S(\tau(m)) > \lambda T$, then (4-9) implies $\frac{S_1(\tau(m))}{\lambda} \geq \frac{\mu_1}{2\lambda} S(\tau(m)) \geq \frac{\mu_1}{2} T > b$ contradicting the fact that $S_1(\tau(m)) \leq b\lambda$). Put

$$S_i(\tau(m+\theta_\lambda-1) - S_i(\tau(m)) = \ell_i$$

for all $i = 1,2,3,\ldots,n$, and

$$S(\tau(m+\rho_\lambda - 1)) - S(\tau(m)) = \ell,$$

then it follows from (4-8), (4-10) that

$$\text{(4-11)} \qquad \lim_{\lambda \to \infty} \left| \frac{\lambda}{\ell} \int_{\frac{S(\tau(m))}{\lambda}}^{\frac{S(\tau(m))}{\lambda}+\frac{\ell}{\lambda}} \delta_i(Y(\lambda s,\omega))ds - \mu_i \right| = 0$$

uniformly on the set $\ell/\lambda \geq \epsilon(\lambda)$.

The lemma follows from the facts that the integral in (4-11) is equal to $\frac{\ell_i}{\lambda}$ and $\epsilon(\lambda) \to 0$ as $n \to \infty$.

Note that $O(\frac{1}{\lambda})$ tends to zero uniformly over all $I \subseteq [0,t]$ since we can take $T = \frac{2t}{\mu_1} + T'$ for all $I = [a,b] \subseteq [0,t]$.

Remark: The existence of the function $\epsilon(\lambda)$ was first proved by Kurtz [6] in a form which is slightly weaker than (4-8) in the sense that our limit is taken over the supremum of all $\epsilon \geq \epsilon(\lambda)$ and $t \leq T$.

Proof of theorem 3-6: Let $\omega \in M$, $\omega^* \in \Omega'$, $t \geq 0$ and $\epsilon > 0$ be fixed. We set $H = \{s \leq t \mid a_1(s) = 0\}$ and $\mu_0 = \min\{\mu_1,\mu_2,\ldots,\mu_n\}$. For each positive integer N, define the set

$$F_{N,0} = \left\{ s \leq t \,\middle|\, \frac{a_1(s)}{a_i(s)} < \frac{1}{N} \text{ for some } i = 2,3,\ldots,n \right\}.$$

If $s \in \bigcap_{N=1}^{\infty} F_{N,0}$ then $a_1(s) \leq \frac{1}{\mu_0 N}$ for all N, therefore $F_{N,0} \downarrow H$ as $N \to \infty$. Since $A_1(t) \equiv t$, the lebesgue measure of H is zero. We choose an N so large that $B(F_{N,0} - H) < \epsilon\mu_0$. Having fixed such an N we drop the subscript N in our notation and write $F_{N,0}$ as F_0.

Let $\mathcal{J} = \{2,3,\ldots,N^2\}$ and $\mathcal{J}^{(n-1)} = \mathcal{J} \times \mathcal{J} \times \ldots \times \mathcal{J}$ (n-1 times). Let $\vec{k}(1), \ldots, \vec{k}((N^2-1)^{n-1})$ be an enumeration of the elements in $\mathcal{J}^{(n-1)}$. For each $\vec{k}(m)$, $m = 1,\ldots,(N^2-1)^{n-1}$, define

$$F_m = F_{N,m} = \left\{ s \leq t \,\middle|\, \frac{k_i - 1}{N} \leq \frac{a_1(s)}{a_i(s)} < \frac{k_i}{N},\ i = 2,\ldots,n, \text{ where } \vec{k}(m) = (k_2,k_3,\ldots,k_n) \right\}$$

and

$$F = \left\{ s \le t \,\middle|\, \frac{a_1(s)}{a_i(s)} \ge N \text{ for some } i = 2,3,\ldots,n \right\}.$$

Then

$$(4\text{-}13) \qquad |C_\lambda(t)-A(t)| \le |C_\lambda(F_0)-A(F_0)|+|C_\lambda(F)-A(F)|+\sum_{i=1}^{(N^2-1)^{n-1}} |C_\lambda(F_i)-A(F_i)|.$$

We now estimate each term in the right hand side of the inequality (4-13). We use ℓ to denote the lebesgue measure on R^1.

We cover H with a sequence $\{I_i\}$ of open intervals with $\sum_{i=1}^{n} \ell(I_i) < \epsilon\mu_1$ and $B(\cup\{I_j\} - H) < \epsilon$ and then choose a Q_0 so large that $B(\cup\{I_j | j > Q_0\}) < \epsilon\mu_0$. Now consider $C_\lambda(I_j)$; it follows from the fact

$$\sum_{i=1}^{n} C_\lambda(G_\lambda^i \cap I) = C_\lambda(I) + O(\tfrac{1}{\lambda})$$

and (ii) of Remark 4-7 that

$$C_\lambda(I_j) \le \frac{\ell(I_j)}{\mu_1} + O(\tfrac{1}{\lambda}).$$

Thus

$$C_\lambda(F_0) < \frac{B(F_0-H)}{\mu_0} + \frac{B(\bigcup_{j\ge Q_0} I_j)}{\mu_0} + \sum_{j=1}^{Q_0} C_\lambda(I_j)$$

$$\le 2\epsilon + \sum_{j=1}^{Q_0} \frac{\ell(I_j)}{\mu_1} + Q_0\cdot O(\tfrac{1}{\lambda}) \le 3\epsilon + Q_0\cdot O(\tfrac{1}{\lambda}).$$

The fact that $O(\frac{1}{\lambda})$ tends to zero uniformly over all $I \subseteq [0,t]$ guarantees the existence of a positive number W_0 such that

$$C_\lambda(F_0) \le 4\epsilon$$

if $\lambda > W_0$. Now the definition of A implies

$$0 \le A(F_0) = A(F_0 - H) \le B(F_0 - H) < \epsilon\mu_0 < \epsilon.$$

Therefore

(4-14) $\quad |C_\lambda(F_0) - A(F_0)| < 5\epsilon$ if $\lambda > W_0$.

To estimate $|C_\lambda(F) - A(F)|$: since $F = \bigcup_{i=2}^{n} \{s \leq t \mid \frac{a_1(s)}{a_i(s)} \geq N\}$, we will estimate $J_i = \{s \leq t \mid \frac{a_1(s)}{a_i(s)} \geq N\}$ for each $i = 2,3,\ldots,n$. Fix an i, we shall drop "i" in our notation and write J instead of J_i.

We cover J with a sequence $\{I_j\}$ of disjoint open intervals with $B(\cup I_j - J) + \ell(\cup I_j - J) < \epsilon\mu_0$ and then choose Q_i such that

$$(\ell + B)(\cup\{I_j \mid j \geq Q_i\}) < \epsilon\mu_0.$$

Then

$$C_\lambda(I_j \cap G_\lambda^i) \leq \int_{I_j \cap G_\lambda^i \cap J} \frac{a_i(s)}{a_1(s)} ds + \frac{B(I_j - J)}{\mu_0}$$

$$< \frac{1}{N}\, \ell(I_j) + \frac{B(I_j - J)}{\mu_0}.$$

Remark 4-7 implies

$$C_\lambda(I_j) = \frac{1}{\mu_i}\, C_\lambda(I_j \cap G_\lambda^i) + O(\tfrac{1}{\lambda}) \leq \frac{1}{\mu_0 N}\, \ell(I_j) + \frac{B(I_j - J)}{\mu_0^2} + O(\tfrac{1}{\lambda}).$$

Thus

$$C_\lambda(J_i) \leq \sum_{j=1}^{Q_i} C_\lambda(I_j) + \frac{B(\cup\{I_j \mid j \geq Q_i\})}{\mu_0}$$

$$\leq \frac{1}{\mu_0 N}\, \Sigma \ell(I_j) + \epsilon + \sum_{i=1}^{Q_i} \frac{B(I_j - J)}{\mu_0} + O(\tfrac{1}{\lambda})$$

$$\leq \frac{1}{\mu_0 N}(\ell(J_i) + \epsilon) + 2\epsilon + O(\tfrac{1}{\lambda})$$

$$\leq \frac{1}{\mu_0 N}\, \ell(J_i) + 3\epsilon + O(\tfrac{1}{\lambda}).$$

Summing over all i, and using the fact that $O(\frac{1}{\lambda})$ tends to zero uniformly over all intervals contained in $[0,t]$, we obtain a number W such that

$$C_\lambda(F) \leq \frac{1}{\mu_0 N}\, \ell(F) + 3n\epsilon + \epsilon \qquad \text{if } \lambda > W.$$

Also it is easy to see that

$$A(F) \le \int_F \frac{1}{\mu_1 + \Sigma \mu_i \frac{a_1(s)}{a_i(s)}}\, ds \le \frac{1}{\mu_0 N}\, \ell(F).$$

Thus

$$(4\text{-}15) \qquad |C_\lambda(F) - A(F)| \le \frac{2}{\mu_0 N}\, \ell(F) + (3n+1)\epsilon \qquad \text{if } \lambda > W.$$

To estimate $C_\lambda(F_k)$ for $k = 1,2,\ldots,(N^2-1)^{n-1}$: We cover F_k with a sequence $\{I_j\}$ of disjoint open intervals with $(\ell + B)(\cup I_j - F_k) < \frac{\epsilon \mu_0{}^2 \mu_1}{N \cdot 2^{k+1}}$ and then choose a Q_k such that $(\ell + B)(\cup\{I_j | j \ge Q_k\}) < \frac{\epsilon \mu_0 \mu_1}{N \cdot 2^{k+1}}$.

Lemma 4-5 and Remark 4-7 imply

$$(*) \quad C_\lambda(I_j \cap G_\lambda^i) = \frac{\mu_i}{\mu_1} C_\lambda(I_j \cap G_\lambda) + O(\tfrac{1}{\lambda})$$

$$= \frac{\mu_i}{\mu_1} \ell(I_j \cap G_\lambda) + O(\tfrac{1}{\lambda}).$$

Also

$$C_\lambda(I_j \cap G_\lambda^i) \le \int_{I_j \cap G_\lambda^i \cap F_k} \frac{a_i(s)}{a_1(s)}\, ds + \frac{B(I_j - F_k)}{\mu_0}$$

$$\le \frac{N}{k_i - 1}\, \ell(I_j \cap G_\lambda^i \cap F_k) + \frac{B(I_j - F_k)}{\mu_0}.$$

Thus

$$\frac{k_i - 1}{N} C_\lambda(I_j \cap G_\lambda^i) \le \ell(I_j \cap G_\lambda^i \cap F_k) + \frac{B(I_j - F_k)}{\mu_0} \frac{k_i - 1}{N}$$

and

$$C_\lambda(I_j \cap G_\lambda^1) = C_\lambda(I_j \cap G_\lambda) = \ell(I_j \cap G_\lambda).$$

Summing over $i = 1,\ldots,n$, we thus obtain (from (*))

$$C_\lambda(I_j \cap G_\lambda) \le \frac{1}{(1 + \sum_{i=2}^{n} \frac{k_i - 1}{N} \frac{\mu_i}{\mu_1})} \ell(I_j) + \frac{\mu_1}{\mu_0^2} B(I_j - F_k) + O(\tfrac{1}{\lambda}).$$

Again Remark 4-7 implies

$$C_\lambda(I_j) \le \frac{1}{(\mu_1 + \sum_{i=2}^{N} \frac{k_i - 1}{N} \mu_i)} \ell(I_j) + \frac{1}{\mu_0^2} B(I_j - F_k) + O(\tfrac{1}{\lambda}).$$

Summing over all $j = 1,\dots,Q_k$, we obtain

$$\sum_{j=1}^{Q_k} C_\lambda(I_j) \le \frac{1}{(\mu_1 + \sum_{i=2}^{N} \frac{k_i - 1}{N} \mu_i)} \sum_{j=1}^{Q_k} \ell(I_j) + \frac{\epsilon}{N \cdot 2^{k+1}} + O(\tfrac{1}{\lambda}).$$

Thus

$$\text{(4-16)} \qquad C_\lambda(F_k) \le \sum_{j=1}^{Q_k} C_\lambda(I_j) + \frac{B(\cup\{I_j \mid j \ge Q_k\})}{\mu_0}$$

$$\le \frac{\sum_{j=1}^{Q_k} \ell(I_j)}{\mu_1 + \sum_{i=2}^{N} \frac{k_i - 1}{N} \mu_i} + \frac{\epsilon}{N2^k} + O(\tfrac{1}{\lambda})$$

$$\le \ell(F_k)\left(\frac{1}{\mu_1 + \sum_{i=2}^{N} \frac{k_i - 1}{N} \mu_i}\right) + \frac{3\epsilon}{\mu_1 N 2^k} + O(\tfrac{1}{\lambda})$$

$$\le \ell(F_k)\left(\frac{1}{\mu_1 + \sum_{i=2}^{N} \frac{k_i}{N} \mu_i}\right) + \ell(F_k)\frac{1}{\mu_1^2 N} + \frac{3\epsilon}{\mu_1 N 2^k} + O(\tfrac{1}{\lambda})$$

$$\le A(F_k) + \ell(F_k)\frac{1}{\mu_1^2 N} + \frac{3\epsilon}{\mu_1 N 2^k} + O(\tfrac{1}{\lambda}).$$

Therefore

$$\overline{\lim_{\lambda \to \infty}}\, C_\lambda(F_k) \le A(F_k) + \frac{1}{\mu_1^2 N} \ell(F_k) + \frac{3\epsilon}{\mu_1 N 2^k}\,.$$

Summing this over $k = 1,2,\dots,(N^2-1)^{n-1}$, and taking into account (4-14) and (4-15), we obtain

$$\overline{\lim_{\lambda \to \infty}}\, C_\lambda(t) \le A(t) + (3n + 5)\epsilon + \frac{1}{N}\left(\frac{2t}{\mu_0 \mu_1^2} + \frac{3\epsilon}{\mu_1}\right).$$

Thus

$$\overline{\lim_{\lambda \to \infty}} C_\lambda(t) \leq A(t).$$

The inequality $\underline{\lim}_{\lambda \to \infty} C_\lambda(t) \geq A(t)$ follows from (4-14), (4-15), and the following inequalities:

$$\begin{aligned}
C_\lambda(F_k) &= C_\lambda(\cup I_j) - C_\lambda(\cup I_j - F_k) \\
&\geq \sum_{j=1}^{Q_k} C_\lambda(I_j) - \frac{B(\cup I_j - F_k)}{\mu_0} \\
&\geq \frac{\Sigma \ell(I_j)}{\mu_1 + \sum_{i=2}^{N} \frac{k_i}{N} \mu_i} - \frac{\epsilon}{N 2^k} + O(\tfrac{1}{\lambda}) \\
&\geq \frac{\ell(F_k)}{\mu_1 + \sum_{i=2}^{N} \frac{k_i - 1}{N} \mu_i} - \frac{\ell(F_k)}{\mu_1^2 N} - \frac{3\epsilon}{N\mu_1 2^k} + O(\tfrac{1}{\lambda}) \\
&\geq A(F_k) - \frac{\ell(F_k)}{\mu_1{}^2 N} - \frac{3\epsilon}{N\mu_1 2^k} + O(\tfrac{1}{\lambda}).
\end{aligned}$$

This proves the theorem.

REFERENCES

[1] Dynkin, E. B., Markov Processes, Vol. I, Springer-Verlag, 1965.

[2] Hersh, R. and Griego, R. J., Random evolution, Markov chains, and systems of P.D.E., Proc. Nat. Acad. Sci. U.S.A., 62(1969), 305-308. MR42 #5099.

[3] Kurtz, T. G., A random Trotter product formula, Proc. AMS 35(1972), 147-154.

[4] Pitt, Loren, Product of Markovian Semigroups of Operators, Z. Wahr. Verw. Geb., 12, 241-254.

Large Deviations for Markov Processes and the Asymptotic Evaluation of Certain Markov Process Expectations for Large Times

M.D. Donsker & S.R.S. Varadhan

We start with a theorem of Kac [5]. Let $\beta(t)$ be one dimensional Brownian motion and $V(x)$ a non-negative continuous function satisfying $V(x) \to +\infty$ as $x \to \pm\infty$. Then

$$\lim_{t\to\infty} \frac{1}{t} \log E_x \left[\exp\left\{-\int_0^t V(\beta(s))ds\right\}\right]$$

exists and equals $-\lambda_1$ where λ_1 is the smallest eigenvalue of

$$\frac{1}{2}\frac{\partial^2\psi}{\partial x^2} - V(x)\psi + \lambda\psi = 0 \quad .$$

Here E_x stands for the expectation with respect to Brownian motion paths starting at time zero from the point x on the line. The idea behind his proof is that under the hypothesis on $V(x)$ we have a complete set of eigenfunctions and eigenvalues and we can write the fundamental solution $P_V(t,x,y)$ of the equation

$$\frac{\partial u}{\partial t} - \frac{1}{2}\frac{\partial^2 u}{\partial x^2} + V(x)u = 0$$

as

$$P_V(t,x,y) = \sum e^{-\lambda_j t}\, \psi_j(x)\psi_j(y)$$

where (λ_j,ψ_j) is a complete set of eigenvalues and corresponding normalized eigenfunctions. The expectation we are interested in is equal to $\int_{-\infty}^{\infty} P_V(t,x,y)dy$ and since

$\lambda_1 < \lambda_2 \leq \lambda_3 \leq \ldots$ and $\psi_1(x) > 0$ for all x, it is clear that the first eigenvalue will dominate.

We also have on the other hand a variational formula for the first eigenvalue λ_1, namely

$$\lambda_1 = \inf_{\psi:||\psi||_2=1} \{ \frac{1}{2} \int [\psi'(x)]^2 dx + \int V(x)\psi^2(x)dx \} .$$

There is apparently no direct reason why an analysis of the asymptotic evaluation of

$$E_x [\exp \{-\int_0^t V(\beta(s))ds\}]$$

should lead to the above variational problem, but as we shall see this direct relation exists and moreover is a special case of a general theory.

There are situations in which asymptotic evaluation of integrals do lead to variational formulae. This is the method of Laplace asymptotic formula. In its simplest situation we have

$$\lim_{t\to\infty} \frac{1}{t} \log \int_0^1 \exp \{- tf(x)\}dx = - \inf_{0\leq x\leq 1} f(x)$$

where f(x) is a continuous function. There are analogues of this in an arbitrary space. An abstract version of this which can be found in [8] is described below.

One has a metric space Ω and a family P_n of probability measures on the Borel sets. It is postulated that there exists a function $I(\omega)$ on Ω having some nice properties such that

$$(1) \qquad P_n(A) \sim \exp \{-n \inf_{\omega\in A} I(\omega)\}$$

then it is proved that for nice functionals $F(\omega)$ on Ω

$$\lim_{n\to\infty} \frac{1}{n} \log \int \exp\{nF_n(\omega)\} P_n(d\omega) = \sup_{\omega} [F(\omega) - I(\omega)]$$

The manner in which (1) is to be satisfied is that

a) for C closed

$$\limsup_{n \to \infty} \frac{1}{n} \log P_n(C) \leq - \inf_{\omega \in C} I(\omega)$$

b) for G open

$$\liminf_{n \to \infty} \frac{1}{n} \log P_n(G) \geq - \inf_{\omega \in G} I(\omega) \quad .$$

$I(\omega)$ is assumed to be a non-negative lower semi-continuous functional on Ω. $F(\omega)$ is assumed to be continuous and bounded below. In [8] we needed a technical assumption that for any $\ell > 0$ the set $\{\omega: I(\omega) \leq \ell\}$ should be a compact subset of Ω. This can be dropped but then it should be replaced by the condition that $\{\omega: F(\omega) \leq \ell\}$ are compact sets for every finite ℓ.

The interpretation for (1) is that if ω_o is such that $I(\omega_o) = 0$ and $I(\omega) > 0$ for $\omega \neq \omega_o$ then P_n converges weakly to δ_{ω_o}. So we can start with a sequence P_n converging weakly to some δ_{ω_o}. Then for any set A separated from ω_o, $P_n(A)$ will tend to zero. We can then ask for the rate of decay of $P_n(A)$ as $n \to \infty$. More precisely we want $P_n(A)$ to go to zero exponentially and we want the exact exponential constant as a function of A. a) and b) which describe (1) formally relate the exponential constant for any set A in terms of the I function. The I function therefore governs the probabilities of large deviations of ω from ω_o. Previous situations where such an approach has proved useful are when a deterministic system is perturbed by a small random term. See for instance [7], [6], [8], [9], and [10].

The current situation is different on the surface because as $t \to \infty$ randomness increases and the measures do not concentrate around a single trajectory. However out of chaos comes order in the form of the ergodic phenomenon.

Let us suppose that $X_1, X_2, \ldots, X_n, \ldots$ is a Markov chain on a finite state space with transition probabilities π_{ij}. We shall assume for simplicity that π_{ij} are all positive. If $f_1^{(n)}, \ldots, f_k^{(n)}$ are the proportions of times the chain spends in the states $1,2,\ldots,k$ respectively before time n then for any starting state x

$$P_x \{|f_1^{(n)} - p_1| < \varepsilon, \ldots, |f_k^{(n)} - p_k| < \varepsilon\} \to 1$$

Let us denote by $Q_{n,x}$ the measure on the set of probability vectors on k – states corresponding to $f_1^{(n)}, \ldots, f_k^{(n)}$ and the starting point x. By the ergodic theorem $Q_{n,x}$ converges to δ_p where p is the invariant probability. One can now ask if there is a functional $I(q)$ on the set of probability vectors such that $I(q)$ is the I function for $Q_{n,x}$ in the sense of (1). The answer is yes and in fact for any $q = \{q_1, \ldots, q_k\}$

$$I(q) = - \inf_{\substack{u_1>0 \\ \cdots \\ u_k>0}} \sum_{j=1}^{k} q_j \log \left(\frac{\sum_{\ell=1}^{k} \pi_{j\ell} u_\ell}{u_j} \right)$$

Then if $V(i)$ is a function on the state space

$$\lim_{n\to\infty} \frac{1}{n} \log E_x [\exp\{V(x_1) +, \ldots, V(x_n)\}]$$

$$= \lim_{n\to\infty} \frac{1}{n} \log E_x [\exp n\{\sum_{i=1}^{k} f_i^{(n)} V(i)\}]$$

$$= \sup_q [\sum_{i=1}^{k} V(i) q_i - I(q)] \quad .$$

This is precisely the discrete analogue of Kac's result for an arbitrary Markov chain where the variational formula comes up naturally.

In the case of the general continuous time version the

role of the proportions $f_1^{(n)},\dots,f_k^{(n)}$ is played by the occupation time

$$L_t(\omega,A) = \frac{1}{t}\int_0^t \chi_A(x(s,\omega))ds$$

Here $x(s,\omega)$ is the Markov process on some state space X and A is a Borel set in X. Then we have the measure $Q_{t,x}$ on the space $\mathcal{M}$ of probability measures on X induced by the map $\omega \to L_t(\omega,\cdot)$ which maps $\Omega \to \mathcal{M}$. Here Ω is the basic function space on which the Markov process induces a family $\{P_x\}$ of measures corresponding to each starting point x and $Q_{t,x}$ is the distribution on $\mathcal{M}$ induced by $L_t(\omega,\cdot)$ from P_x. If μ_0 is the invariant measure for the process then $Q_{t,x}$ will concentrate around μ_0 and the analogue of (1) can be proved with $I(\mu)$ defined to be

$$I(\mu) = -\inf_{u>0}\int\left(\frac{Lu}{u}\right)\mu(dx)$$

where the infinum is taken over all functions u which are bounded, have a positive lower bound and are in the domain of the infinitesimal generator L of the process. Some technical assumptions are needed in the proof and we will not try to describe them here. But instead refer to [2] where the result is proved in some generality.

The Brownian motion case we started out with is still not quite covered because then there is no invariant measure. However this makes only a technical difference in that $I(\mu)$ will now be strictly positive for all μ and never zero. The actual computation for the Brownian motion case yields that $I(\mu)$ is finite if and only if μ has a density with respect to Lebesgue measure and this density has a non-negative square root which has a square integrable derivative. Then

$$I(\mu) = \frac{1}{8}\int \frac{(f')^2}{f}dx = \frac{1}{2}\int[(\sqrt{f})']^2 dx$$

so by changing $\sqrt{f}$ to ψ we pass from the variational formula obtained by Laplace asymptotic method to one that gives the classical formula of Rayleigh and Ritz. This is carried out in [1].

There is also the question of the topology on $\mathcal{M}$. It could be the weak or the norm topology. To get results in the norm topology the problem has to be modified slightly. See [3] where this is carried out.

The advantage now is that the functional $F(\mu)$ does not have to be of the form $\int V(x)\mu(dx)$ which is the case of the potential considered by Kac. We can replace it by any nice functional $F(\mu)$ and we can prove the asymptotic formula

$$\lim_{t\to\infty} \frac{1}{t} \log E_x[\exp t\{F(L_t(\omega,\cdot))\}] = \sup_{\mu} [F(\mu) - I(\mu)]$$

An interesting application of such a case, where the norm topology has to be used, is the solution of the Wiener Sausage problem which is carried out in [4].

References

[1] Donsker, M.D. and Varadhan, S.R.S., Asymptotic Evaluation of Certain Wiener Integrals for large time, Proceeding of the International Conference on Integration in function spaces, Clarendon Press, Oxford, 1974

[2] Donsker, M.D. and Varadhan, S.R.S., Asymptotic Evaluation of certain Markov process expectations for large time, to appear in Comm. Pure Appl. Math.

[3] Donsker, M.D. and Varadhan, S.R.S., Asymptotic Evaluation of certain Markov process expectations for large time II, to appear in Comm. Pure Appl. Math.

[4] Donsker, M.D. and Varadhan, S.R.S., The asymptotics of the Wiener Sausage, to appear in Comm. Pure Appl. Math.

[5] Kac, M., On some connections between probability theory and differential and integral equations, Proceedings of the Second Berkeley Symposium 1950, pp. 189-215

[6] Pincus, M., Gaussian Processes and Hammerstein Integral Equations, Trans. Amer. Math. Soc., 134 (1968) pp. 193-216

[7] Schilder, M., Some Asymptotic formulae for Wiener Integrals, Trans. Amer. Math. Soc., 125 (1966) pp. 63-85

[8] Varadhan, S.R.S., Asymptotic Probabilities and differential equations, Comm. Pure Appl. Math. vol. 19, 1966 pp. 261-286

[9] Varadhan, S.R.S., Diffusion Processes in a small time Integral, Comm. Pure Appl. Math. vol 20, 1967 pp. 659-685

[10] Ventsel, A.D. and Freidlin, M.I., On small Random perturbations of dynamical systems, Russian Mathematical Surveys, vol. 25, no. 1, (1970) pp. 1-55

Random Evolutions

Mark A. Pinsky

In this talk we give a treatment of random evolutions using a stochastic calculus. In this way we can give a unified treatment of multiplicative operator functionals for both diffusion processes and jump processes.

Random evolutions were introduced by Griego and Hersh in 1969 [2]. Their definition is an abstract counterpart of the one-dimensional random velocity model studied by the author. In subsequent works [4,10] it was found that their abstract construction is the Markov-chain analogue of the Feynman-Kac and Cameron-Martin formulas known for the Wiener process. All of these developments are covered in the author's survey [10] and [3].

We begin by formulating the random evolution in terms of a martingale problem, in the spirit of Stroock and Varadhan [11].

1. A Martingale Problem

In order to formulate the connection with martingales, we first review the definition of multiplicative operator functional of a Markov process.

Definition. Let X be a Markov process and L a separable Banach space. A MOF (multiplicative operator functional) is a mapping $(I,\omega) \to M(I,\omega) \in \mathcal{B}(L,L)$ where I is an interval in $[0,\infty)$ and which satisfies the following:

i) $\omega \to M(I,\omega)$ is measurable $F_I(\equiv \sigma\{\xi(s):\ s\in I\})$

ii) $I \to M(I,\omega)$ is strongly continuous a.s. in the following sense: if $I_n \uparrow I$ or $I_n \downarrow I$ when $n \to \infty$, then s-lim $M(I_n,\omega) = M(I,\omega)$ a.s.

iii) $M(I_1 \cup I_2,\omega) = M(I_1,\omega)\, M(I_2,\omega)$ a.s. whenever I_1 and I_2 are intervals such that $I_1 \cup I_2$ is an interval and I_1 lies to the left of I_2.

iv) $M(I + t,\omega) = M(I,\theta_t\omega)$ a.s. for each $t > 0$.

v) $M(\varnothing,\omega) = I$ a.s. ($\varnothing$ = the empty interval, I = identity operator)

The connection between MOF's and differential equations can be formulated in terms of the following <u>martingale problem</u>: given an equation

$$\frac{\partial f}{\partial t} = Gf$$

find a Markov process X and a MOF $M(0,t]$ such that for all smooth functions f

$$M(0,t]f(\xi(t)) - \int_0^t M(0,s]Gf(\xi(s))ds$$

is a local martingale. (We shall suppress the ω-dependence and write $M(0,t]$ if $I = (0,t]$).

<u>Example 1</u>. If X is a right-continuous Markov process and V is a non-negative Borel function, $L = R^1$, we may take $M(0,t] = \exp(- \int_0^t V(\xi(s))ds$. This corresponds to the Feynman-Kac formula.

Example 2. If $\{\xi(t), t \geq 0\}$ is the Brownian motion on a Riemannian manifold of dimension d and $L = R^d$, then we may take $M(0,t]$ = stochastic parallel transport of a vector f along the trajectory ω (c.f. [5] and the talks of K. Itô and P. Malliavin in this volume).

Remark. If the local martingale above has finite expectation and $M(0,t]$ has finite expectation, then the mapping $f \to E\{M(0,t] f(\xi(t))\}$ is a semigroup of linear operators, whose infinitesimal generator is an extension of G acting on smooth functions. The solution of the equation $f_t = Gf$ has the representation $f(t,\xi) = E_\xi\{M(0,t]f(0,\xi(t))\}$.

2. A Stochastic Calculus

We will illustrate the martingale problem in two important cases:

Case 1. $\xi(t)$ is the d-dimensional Wiener process, $L = R^N$ where d,N are arbitrary natural numbers. By a known result [9] any smooth MOF can be obtained as the solution of the linear Itô equation

$$(1) \quad M(0,t] = I + \int_0^t M(0,s]B_0(\xi(s))ds + \sum_{j=1}^{d} \int_0^t M(0,s]B_j(\xi(s))d\xi_j(s).$$

The final term is an Itô stochastic integral. On the other hand, by Itô's formula, for any $f \in C^2$

$$(2) \quad f(\xi(t)) = f(\xi(0)) + 1/2 \int_0^t \Delta f(\xi(s))ds + \sum_{j=1}^{d} \int_0^t \frac{\partial f}{\partial \xi_j} d\xi_j(s)$$

Now we use the stochastic product rule $d(Mf) = Mdf + (dM)f + (dM)(df)$. After rearranging terms, we get

$$(3)\quad M(0,t]f(\xi(t)) = f(\xi(0)) + \int_0^t M(0,s]\{\tfrac{1}{2}\Delta f + \sum_{j=1}^{d} B_j \frac{\partial f}{\partial \xi_j} + B_0 f\}(\xi(s))ds + Z_t$$

where $Z_t = \sum_{j=1}^{d} \int_0^t M(0,s]\{\frac{\partial f}{\partial \xi_j} d\xi_j(s) + B_j(\xi(s))f(\xi(s))d\xi_j(s)\}$. Z_t is an Itô stochastic integral and hence a local martingale. Therefore we can make the identification

$$(4)\qquad Gf = 1/2\ \Delta f + \sum_{j=1}^{d} B_j \frac{\partial f}{\partial \xi_j} + B_0 f.$$

<u>Case II</u>. Let $\{\xi(t),\ t \geq 0\}$ be a conservative regular jump process on a state space E. Such a process has jump times $0 < \tau_1 < \tau_2 < \ldots \to \infty$. The "road map" of $\{\xi(\tau_1),\ \xi(\tau_2),\ldots\}$ is governed by a transition kernel $\pi(\xi, d\eta)$; the time scale is determined by a positive function $\nu(\xi)$. Let L be an arbitrary separable Banach space.

It has been shown that any MOF of norm not greater than one is determined by the following rules [10]:

$$(5)\qquad M(0,\tau_1) = \exp(\tau_1 A_{\xi(0)})$$

$$(6)\qquad M[\tau_1,\tau_1] = B_{\xi(\tau_1),\xi(\tau_1)}$$

where A_ξ is the generator of a strongly continuous contraction semigroup and $B_{\xi\eta}$ is a contraction operator. Using the multiplicative property, we have

(7)

$$M(0,t] = \left[\prod_{i=1}^{N(t)} \exp((\tau_i-\tau_{i-1})A_{\xi(\tau_{i-1})})B_{\xi(\tau_{i-1}),\xi(\tau_i)}\right] \exp\{(t-\tau_{N(t)})A_{\xi_{N(t)}}\}$$

where N(t) is the integer k which satisfies $\tau_k \leq t < \tau_{k+1}$.

Between jump times, M evolves according to the equation $dM/dt = MA_{\xi(t)}$. Thus we have the equation

$$(8) \quad M(0,t] = I + \int_0^t M(0,s]A_{\xi(s)}ds + \sum_{s\leq t}{}' M(0,s)[B_{\xi(s^-),\xi(s)} - I]$$

where Σ' denotes the sum over all jump times $\tau_j \leq t$. On the other hand, for any Borel function f we have the telescoping sum

$$(9) \qquad f(\xi(t)) = f(\xi(0)) + \sum_{s\leq t}{}' [f(\xi(s)) - f(\xi(s^-))].$$

Now we use the stochastic product rule in the form $d(Mf) = M^-(df) + (dM)f^- + (dM)(df)$ where we must use the indicated left-hand limits. (This equation can be proved by inspection for these jump processes; it also follows from the general theory [8]). The result is

$$(10) \quad M(0,t]f(\xi(t)) = f(\xi(0)) + \int_0^t M(0,s]A_{\xi(s)}f(\xi(s))ds$$

$$+ \sum_{s\leq t}{}' M(0,s)[B_{\xi(s^-),\xi(s)}f(\xi(s)) - f(\xi(s^-))]$$

Clearly equations (8-10) play the same role for the jump process that equations (1-3) play for the Wiener process. But in this case the last term on the right side of (10) is not a local martingale. In order to find the local martingale we use the following equation: for any positive Borel function $\varphi(.,.)$,

$$(11) \qquad E\{\sum_{s\le t}{}' \varphi(\xi(s^-),\xi(s))\} = E\{\int_0^t \nu(\xi(s))(\pi\varphi)(\xi(s))ds\}$$

where $(\pi\varphi)(\xi) = \int_E \varphi(\xi,\eta)\pi(\xi,d\eta)$. This is a special case of the Levy system [12] for a Markov process. Applying (11) to (10), we get

$$M(0,t]f(\xi(t)) = f(\xi(0)) + \int_0^t M(0,s)Gf(\xi(s))ds + Z_t$$

where

$$(12) \qquad Gf(\xi) = A_\xi f + \nu(\xi)\int_E [B_{\xi\eta}f(\eta) - f(\xi)]\pi(\xi,d\eta),$$

and Z_t is a local martingale.

3. Two examples

For the remainder of this talk $\xi(t)$ is a regular jump process. We will specialize the form (7) to obtain two models of independent interest.

Example 1:(Random flow) Let $\xi(t)$ be a finite-valued Markov process taking values in $\{1,\dots,N\}$. Let $v_1(x),\dots,v_N(x)$ be vector fields on R^d, $L = C_o(R^d)$, and set $A_\xi = v_\xi\cdot\nabla$, $B_{\xi\eta} = I$. This model is equivalent to studying the solution $X^x(t)$ of the random ordinary differential equation

$$(13) \qquad dx/dt = v_{\xi(t)}(x) \quad , X^x(0) = x\in R^d$$

In fact $M(0,t]f = f(X^x(t))$; the corresponding evolution equation for functions is

$$\frac{\partial f}{\partial t} = v_\xi . \nabla f + \sum_{\eta=1}^{N} q_{\xi\eta} f(\eta) \tag{14}$$

where $(q_{\xi\eta})$ is the infinitesimal matrix for the process $\xi(t)$. This model is a continuous, piecewise smooth motion along the flow lines of the vector fields $\{v_1, \ldots, v_N\}$. The construction also makes sense if we replace R^d by a smooth manifold.

Example 2:(Linearized Boltzmann equation) In this case we let $\xi(t)$ be a Markov process on R^3 corresponding to the jump measure $\pi(\xi, d\eta) = (2\pi)^{-3/2} e^{-\eta^2/2} d\eta$ and an otherwise arbitrary function $\nu(\xi)$; $L = C_o(R^3)$, $A_\xi f = -\xi \cdot \nabla f$, $B_{\xi\eta}$ = multiplication by a function $k(\xi,\eta)$. Following the general development in section 2 above, equation (12) tells us that the equation $f_t = Gf$ must be

$$\frac{\partial f}{\partial t} + \xi \cdot \nabla f = \nu(\xi) \int_{R^3} [k(\xi,\eta) f(\eta) - f(\xi)] \frac{e^{-\eta^2/2}}{(2\pi)^{3/2}} d\eta. \tag{15}$$

This example contains as a special case the linearized Boltzmann equation with finite cross section [1]. In this case, the kernel $\nu(\xi)k(\xi,\eta)$ is symmetric and defines a compact operator on the Hilbert space $H \equiv L^2(R^3, e^{-\eta^2/2} d\eta)$. By classical manipulations used to prove

Boltzmann's H-theorem (see, e.g. [1]), it can be shown that the integral operator in (15) is negative semi-definite. From this it is not difficult to show that equation (15) gives rise to a contraction semigroup on H. However it seems difficult to conclude from this the stochastic representation

$$\text{"}f(t,\xi) = E\{M(0,t] f(0,\xi(t))\}\text{"}$$

since we can not show that $M(0,t]$ defined by (7), has finite expectation.

4. Limit theorems

It is of great interest to study the diffusion limit of random evolution. For the case of the linearized Boltzmann equation, we refer to the talk of R. Ellis in this volume. We shall outline below a general operator approach which applies to the random flow model.

Let $\xi(t)$ be an irreducible Markov process with state space $\{1,\ldots,N\}$. The invariant distribution is π_ξ and $C_{\xi\eta} = \pi_\xi \int_0^\infty [p_{\xi\eta}(t) - \pi_\eta]dt$. Consider a one-parameter family of MOF's $M_\epsilon(0,t]$ defined by equation (7), where we assume an asymptotic development

$$A_\xi = \epsilon A_\xi^{(1)} + \epsilon^2 A_\xi^{(2)} + o(\epsilon^2) \qquad (\epsilon \to 0)$$

$$B_{\xi\eta} = I + \epsilon B_{\xi\eta}^{(1)} + \epsilon^2 B_{\xi\eta}^{(2)} + o(\epsilon^2) \qquad (\epsilon \to 0)$$

and where $o(\epsilon^2)$ is in the strong sense for $f \epsilon D$, where D is a dense set in L which is independent of (ξ,η). The following two theorems correspond, respectively, to the weak law of large numbers and the central limit theorem.

Theorem 1. If $V = \sum_{\xi=1}^{N} \pi_\xi A_\xi^{(1)} + \sum_{\xi,\eta} \pi_\xi q_{\xi\eta} B_{\xi\eta}^{(1)}$ generates a semigroup e^{tV}, then

$$\lim_{\epsilon\to 0} E_\xi\{M_\epsilon(0,t/\epsilon]f(\xi(t/\epsilon))\} = e^{tV}\left(\sum_{\xi=1}^{N} \pi_\xi f(\xi)\right)$$

Theorem 2. If $V = 0$, then under supplementary smoothness conditions (cf. Kertz [6]).

$$\lim_{\varepsilon\to 0} E_\xi\{M_\varepsilon(0,t/\varepsilon^2] f(\xi(t/\varepsilon^2))\} = e^{tW}\Big(\sum_{\xi=1}^{N} \pi_\xi f(\xi)\Big).$$

where W is explicitly obtained in terms of $A_\xi^{(1)}$, $A_\xi^{(2)}$, $B_{\xi\eta}^{(1)}$, $B_{\xi\eta}^{(2)}$. The proof of these two results in the above generality is due to R. Kertz [6] who adapted the semigroup method of T. G. Kurtz [7]. In the special case where $A_\xi^{(2)} = 0 = B_{\xi\eta}^{(1)} = B_{\xi\eta}^{(2)}$, the operator W has the form

$$W = \sum_{\xi,\eta=1}^{N} C_{\xi\eta} A_\xi^{(1)} A_\eta^{(1)}$$

which was first proved by Hersh and Papanicolaou [4]. In the general case, Kertz obtained an explicit expression for W which involves terms which are linear in $A_\xi^{(2)}$, $B_{\xi\eta}^{(2)}$ plus terms which are quadratic in $A_\xi^{(1)}$, $B_{\xi\eta}^{(1)}$.

As an application, we consider the scaled random flow

$$\frac{dx_\varepsilon}{dt} = \varepsilon^{1+\delta(\xi(t))} v_{\xi(t)}(x)$$

where $\delta(\xi) = 1$ for $\xi = 0$, $\delta(\xi) = 0$ for $\xi \neq 0$; $\xi(t)$ is a Markov process with state space $\{-r,\dots,r\}$ and $v_{-r},\dots,v_r$ are vector fields such that $v_{-\xi} = -v_\xi$. The process $x_\varepsilon(t)$ can move forward or backward along v_ξ for $\xi \neq 0$ but <u>only forward</u> along v_0. For definiteness, we assume that $\xi(t)$ only makes transitions of the form $\xi \to \xi + 1$ or $r \to -r$. If we apply the above limit theorems, we find that $V = 0$ and

$$W = \text{const.}\ \Big(v_0\cdot\nabla + \sum_{j=1}^{r} (v_j\cdot\nabla)^2\Big)$$

If v_j is uniformly Lipschitzian, this is the generator of a diffusion process on R^d. But any degenerate elliptic operator can be written in this form provided that the matrix of leading coefficients has a smooth square root. Thus we see that for a pre-assigned operator of this type, the associated diffusion process is the weak limit of a sequence of random flows, i.e.

$$\exp(tW)f = \lim_{\epsilon \to 0} E\{f(x_\epsilon(t/\epsilon^2))\}$$

This has the following consequence: if we write $\exp(tW)f = \int p(t,x,dy)f(y)$, then the support of the measure $p(t,x,.)$ must be contained in the set of points which can be reached, starting from x, by continuous paths which are piecewise identical to solutions of the ordinary differential equations $dx/dt = \pm v_\xi(x)\ (\xi \neq 0)$ or $dx/dt = v_0(x)$. By a different approximation of the diffusion process, Stroock and Varadhan [11] were able to precisely characterize the support of the diffusion process in terms of solutions to ordinary differential equations.

References

1. H. Grad, Asymptotic theory of the Boltzmann equation I. Proceedings of the Conference on Rarefied Gas Dynamics, vol. 1, 26-60, Academic Press, New York, 1963.

2. R. Griego and R. Hersh, Random evolutions, Markov chains, and systems of partial differential equations, Proc. Nat. Acad. Sci, 62, 1969, 305-308.

3. R. Hersh, Random Evolutions: A survey of results and problems, Rocky Mountain Journal of Mathematics 4, 443-477, 1974.

4. R. Hersh and G. Papanicolaou, Non-commuting random evolutions and an operator-valued Feynman-Kac formula, Comm. Pure Appl. Math 25, 337-366 (1972).

5. K. Itô, The Brownian motion and tensor fields on Riemannian manifolds, Proceedings of the International Congress of Mathematicians, Stockholm, 1962, 536-540.

6. R. Kertz, Perturbed Semigroup limit theorems with applications to discontinuous random evolutions, Trans. Amer. Math. Soc. 199, 29-54, 1974.

7. T. G. Kurtz, A limit theorem for perturbed operator semigroups with applications to random evolutions. J. Functional Analysis, 12, 1973, 55-67.

8. P. A. Meyer, Seminaire de Probabilitiés I, Springer-Verlag Lecture Notes in Mathematics, vol. 39, 1967.

9. M. Pinsky, Stochastic integral representations of multiplicative operator functionals of a Wiener process, Trans. Amer. Math. Soc. 167, 89-104, 1972.

10. M. Pinsky, Multiplicative operator functionals and their asymptotic properties, Advances in Probability III, 1974 Marcel Dekker, New York.

11. D. Stroock and S. Varadhan, On the support of diffusion processes, with application to the strong maximum principle, Sixth Berkeley Symposium on Mathematical Statistics and Probability, 1970.

12. S. Watanabe, On discontinuous additive functionals and Levy measures of a Markov process, Japanese Journal of Mathematics, 34, 53-70, 1964.

AN APPLICATION OF BRANCHING RANDOM FIELDS TO GENETICS

Stanley Sawyer*

1. Background. The purpose here is to use the theory of branching processes to illuminate a problem in genetics which has led to some highly singular partial differential equations in the past.

Assume we have a population of some kind of creature in a domain D. The individuals in the population move around in D independently according to some migration law, and are also subject to mutations, deaths, and giving birth. The mutations give rise to slightly different types of creatures; we assume that all of these interbreed, and have the same migration and birth-death behavior as the original. In particular, the mutations are selectively neutral; i.e. none of the mutant subtypes have a selective or reproduction advantage over any of the others.

An outstanding problem in descriptive biology is to explain the tremendous diversity in nature. Anthropologists, for example, measure a great number of traits which distinguish human populations (average height, average sitting height, ratio of sitting height to standing height, ratio of head length to head width, and voluminous blood and immunological variations) not all of which can be easily explained as environmental. One school of biologists says that most such variations are due to a kind of genetic random walk, with selective pressure causing changes at a rate which is not much faster. The classical theory holds that most such variations are Darwinian; i.e. are the result of natural selection.

As time progresses, in the model with selectively neutral mutations described above, individuals which are close to one another will be more likely to be of the same genetic type than individuals located farther

*--This research was partially supported by the National Science Foundation under grant NSF GP-21063.

apart. The problem is to distinguish this random variation from the variation caused by selective pressure depending on position. For example, there might be, in one interbreeding species, a selective advantage for one genetic type in one location and for another type at another location. The resulting situation is called a <u>cline</u>. A typical example of a cline is the distribution of a particular kind of mouse in Alabama, which tends to be brown away from the Gulf of Mexico and to be sandy-colored further south (Moran (1962)). Here predators such as owls apparently give a selective advantage to brown mice in forest areas and to sandy-colored ones near the beach. Another example is Bergmann's law, which says that individuals of a species tend to be larger in temperate zones and smaller in tropic or arctic regions. Weight distributions of both American Indians and of mountain lions are examples of this (Stewart (1973)).

For discrete time and space, the expected variation of genetic type with distance for selectively neutral mutations was calculated by Malécot (1948) and independently by Weiss and Kimura (1964-65). Both approaches extend earlier work of Sewall Wright. Assuming D discrete amounts to assuming an array of colonies in D. The individuals in the population migrate between colonies, undergo mutation, and mate within each colony in some order in each generation. The resulting formulas were applied to the distribution of different types of shell configurations in a particular kind of snail (Cepea nemoralis); the distribution was found to be consistent with selective neutrality (Malécot (1948)). See Kimura and Ohta (1971) for a discussion of the Weiss-Kimura results and their applications.

There is not too much trouble extending these results to either time or space being continuous. However, if both time and space are continuous, for example if the migration law is Brownian motion in R^d, the resulting equations for the expected genetic variation become highly singular and have no non-trivial solutions for $d > 1$ (Fleming and Su (1973-74), Nagylaki (1974); see equation (6.4) below). Another approach, based on viewing genetic variability as a diffusion process in $L^2(D)$, also leads to singularities for $d > 1$ (Fleming (1974a), Dawson (1972,1974)).

Our approach here is to avoid these difficulties by considering a class of rare mutant genetic types. This leads to a branching process approximation for the migration model (see §4). Since the model is then well-defined, the corresponding expected variation with distance can be written down, and equations for it derived as an afterthought.

2. Genetics: a brief introduction. For definiteness, assume we are dealing with a diploid dioecious creature such as man. Dioecious means there are two sexes, and all mating is intersexual. Diploid means that every individual's genetic information is carried on a number of pairs of chromosomes (23 pairs in man); during mating each parent contributes one chromosome from each pair to each offspring (see below).

An atom of genetic information on a chromosome is called a gene. Physically, it appears to be a piece of a chromosome which controls the production of one enzyme. Thus if there are two possibilities for genes (say A and B) at one location of a chromosome, there will be four possibilities on two chromosomes, AA, AB, BA, and BB. Since AB and BA almost always have the same effect, they are identified, making three genetic types or genotypes.

In many situations, A controls the production of a particular enzyme, and B either has no effect or produces a defective enzyme. If one active gene can produce enough enzyme for the body's needs, then the genotypes AA and AB will have the same effect on the individual, and BB some other effect. In this case A is called dominant and B recessive; examples are many serious genetic diseases or metabolic defects such as sickle cell anemia or phenylketonuria. Usually, however, the effect of the AB is intermediate.

When individuals of opposite sex mate, the chromosome pairs of the parents separate, and each offspring gets a copy of one of the chromosomes from each pair of chromosomes of each parent. The four possibilities for each of the offspring's chromosome pairs are equally likely. Thus an ABxAB pairing (i.e., both parents are AB at some location on some chromosome) will produce an AA offspring 25% of the time, an AB 50%, and a BB 25%.

By random mating within a colony of size 2N (N males and N females) we mean that 2N matings take place with one offspring each, producing N males and N females. The sets of parents involved is the result of 2N independent choices from the N^2 possible male-female pairs; particular males or females may take part in more than one mating.

The assumptions underlying random mating seem to be that every possible male-female pair produces offspring independently of every other pair, with the distribution of the number of offspring being Poisson with the same mean. If we condition on a total of 2N offspring, "random mating" results. These assumptions do have advantages; for example, the distribution of genotypes of the offspring can be found by chosing a gene at random from the pool of 2N genes of the fathers, and then (independently) one from the maternal gene pool. For different individuals, the choices are independent. This in fact characterizes random mating.

In general, the ancestry of a particular gene can be followed back through several generations, since each gene has only one parent gene at each stage. For two separate genes, either all ancestors are distinct, or the ancestors all coincide before some first common ancestor. In the first case, random mating implies that the first-generation ancestors of the two genes are chosen independently and at random from the initial gene pool. In the second case, assuming there are no mutations, the genes are called identical by descent (i.b.d.). If there are mutations, the genes are i.b.d. only if there are no intervening mutations between these genes and their first common ancestor.

3. Measuring genetic variation. The most natural way of measuring expected genetic variation with distance is the following. Assume there are two types of genes in the population, A and B. In the n-th generation, choose a gene at random from the individuals at a point P in D, and another (independently) at Q in D. Then we can consider the correlation coefficient $K_n(P,Q)$ of the events that one of these genes or the other is an A. Another approach, which is most reasonable if there is an infinite number of genes but which can always be used, is to student the probability $I_n(P,Q)$ that these two genes are identical by descent (see §2).

Curiously enough, if the initial colonies are large and have the same proportion p of A genes, and there is no mutation, $K_n(P,Q) \equiv I_n(P,Q)$ For, random mating implies the expected proportion of A's in each colony is always p, so $K_n = [\text{Prob(both are A)} - p^2]/p(1-p)$. However $r = \text{Prob(both are A)} = I_n \text{Prob(both A} \mid \text{they are i.b.d.)} + (1-I_n)$ $\text{Prob(both are A} \mid \text{not i.b.d.)}$. By considering the distribution of the first-generation ancestors of the two genes as in §2, we conclude $r = I_n p + (1-I_n)p^2 = I_n p(1-p) + p^2$ and $K_n = I_n$. (This is a standard result.)

In the presence of mutations, K_n is what I_n would be at twice the mutation rate (assuming the same mutation rate u of A to B and B to A).

and $p = \frac{1}{2}$). For, assume the first common ancestor of the two genes is M generations back, setting $M = \infty$ if there is no common ancestor. Then $I_n = \text{Exp}((1-u)^{2M})$, since the genes are i.b.d. iff there are no intervening mutations. Similarly $K_n = (r - \frac{1}{4})/\frac{1}{4}$ as before ($p = \frac{1}{2}$), where, given M, $r = \frac{1}{2}\sum\binom{2M}{2n}(1-u)^{2n}u^{2M-2n}$ (= the probability of an <u>even</u> number of mutations in both lines of descent) $= \frac{1}{2}((1-u+u)^{2M} + (1-u-u)^{2M})/2 = \frac{1}{4}(1 + (1-2u)^{2M})$, and $K_n = \text{Exp}((1-2u)^{2M}) = I_n(2u)$, qed. The above analyses are only valid for $n \geq 2$ and $P \neq Q$; otherwise we have to distinguish between two genes in one individual at P and two genes in different individuals, and the situation becomes more complex.

In the following, we will restrict ourselves to working with I_n, since the results will of course be almost the same.

<u>4. A branching process approximation.</u> Assume there are two types of genes (A and B) and hence three genotypes, AA, AB, and BB, in a population of N males and N females. Let the population undergo random mating as in §2, and let Z(n) be the total number of A genes in the population in the n-th generation. If N becomes infinite but Z(n) remains finite, then Z(n) converges to a branching process (Moran (1962), Crow and Kimura (1970)). This is partly because the A genes become so rare that two AB's never meet. Hence all matings are either ABxBB or BBxBB, and by a standard limit theorem for binomial sampling

$$(4.1) \qquad \text{Prob}(Z(n+1) = m \mid Z(n) = k) = e^{-sk}(sk)^m/m!$$

where s is the selective advantage of AB with respect to BB. In particular Z(n) is a branching process with Poisson offspring distribution.

If we have two types of genes in a migrating population, where the population density is very large but the number of A genes remains locally finite, the resulting number distribution of A genes is a multi-type branching process with position in D as type. If time is continuous, the number distribution process is a branching diffusion process in the sense of Ikeda-Nagasawa-Watanabe (1968-69) or Sawyer (1970). In particular, when an AB individual mates (or branches) at a time T and a point a in D, then as in (4.1)

$$(4.2) \qquad \mathrm{Prob}(N_a(T+) = m \mid N_a(T-) = 1) \quad = \quad e^{-s(a)} s(a)^m/m!$$

where $N_a(t)$ is the number of AB individuals at a at time t, and s(a) is the selective advantage of AB at a.

In the following, we consider various subtypes of the A gene (i.e., $A_1, A_2, \ldots, A_n, \ldots$) and look for the probability of i.b.d. for the A_i genes in the various A_iB genotype individuals.

5. Basic results. For definiteness, assume the migration law has a transition density p(t,a,x) with respect to a measure dx in D, and s(a) in (4.2) is constant. Assume every AB individual in the population branches or breeds in any time interval (s,s + dt) with probability Vdt for some positive constant V. As implied above, we keep track of the AB individuals only, ignoring the background BB's; if there is a B to A mutation rate we handle it as a branching process with immigration (see Sawyer (1974)). Finally, let $N_E(t)$ be the (random) number of AB individuals in the set E in D at time t. Then

Theorem 1. Assume that initially there was only one A gene, located in an individual at the point a in D. Then

$$\mathrm{Exp}(N_E(t)) \quad = \quad \int_E m(t,a,x)\,dx,$$

where $m(t,a,x) = e^{mt} p(t,a,x)$ for $m = V(s-1)$; i.e.

$$(5.1) \qquad m(t,a,x) = e^{mt} p(t,a,x), \qquad m = V(s-1).$$

Thus if $N_y(t)$ is the number of individuals at y as in (4.2)

$$\mathrm{Exp}(N_y(t)) = m(t,a,y)dy = e^{mt}p(t,a,y)dy, \quad m = V(s-1).$$

<u>Theorem 2.</u> If initially there is one A gene at a,

$$\mathrm{Exp}\left[N_E(t)N_F(t) - N_{E\cap F}(t)\right] = \int_E \int_F v(t,a,x,y)\,dxdy$$

for all sets E,F in D, where

$$(5.2) \qquad v(t,a,x,y) = Vs^2 \int_0^t \int_D m(t-s,a,b)m(s,b,x)m(s,b,y)\,dbds.$$

These formulas were first derived by Ikeda-Nagasawa-Watanabe (1969-III,p143); for alternate proofs see Sawyer (1974,Appendix).

<u>Remark:</u> If the measure dx is continuous and $|E| = \int_E dy$, then $\mathrm{Prob}(N_E(t) \geq 2) \leq \frac{1}{2}\mathrm{Exp}\left[N_E(t)(N_E(t)-1)\right] = \sigma(|E|)$ as $E \downarrow x$ by Theorem 2, and $\lim \mathrm{Exp}(N_E(t))/|E| = \lim \mathrm{Prob}\left[N_E(t)=1\right]/|E| = m(t,a,x)$. Hence $m(t,a,y)dy$ also measures the probability of one individual at y; i.e.

$$(5.3) \qquad \mathrm{Prob}(N_y(t) = 1) = m(t,a,y)dy.$$

Alternately, it follows from results in Sawyer (1970) that $N_y(t)$ can never be bigger than one. Similarly, if $x \neq y$,

$$(5.4) \qquad \mathrm{Prob}(N_x(t) = 1 \text{ and } N_y(t) = 1) = v(t,a,x,y)dxdy.$$

Thus (5.2) can be viewed as an integral over the possible mating times and positions of the first common ancestor of the genes which are at x and y at time t. Indeed, $s^2 = \sum n(n-1)p_n$ for $p_n = e^{-s}s^n/n!$ is the expected number of <u>pairs</u> of AB individuals produced by this branch, and $m(s,a,b)db\,Vds$ turns out to be the probability of there being a branch (birth) in dbds (Sawyer (1970,1974)).

Theorem 3. Assume dy is continuous, and that all A genes are subject to a spontaneous mutation rate of u. Then, if initially there is one A gene at a in D,

(5.5) $v_u(t,a,x,y)$ = Prob(There exists AB individuals at x and y which are i.b.d.)/dxdy

$$= Vs^2 \int_0^t \int_D m(t-s,a,b)e^{-2us}m(s,b,x)m(s,b,y)\, dbds.$$

Proof: See Sawyer (1974). Note that the formula follows heuristically from the remark, since after the first common ancestor, all mutations should be counted as deaths.

Corollary (Sawyer(1974)): Let L be the infinitesimal generator of the semi-group $T_t f(a) = \int p(t,a,x)f(x)dx$, and L^* its adjoint under $(f,g) = \int_D f(x)g(x)dx$. Then $v_u(t,a,x,y)$ is a weak solution of

$$(5.6) \quad \frac{\partial}{\partial t} v_u = (L_x^* + L_y^* + 2m - 2u)v_u + Vs^2 e^{mt}\, p(t,a,x)\,\delta(x-y).$$

6. An initial field of particles. Assume that, instead of there being initially only one A gene, there is initially a random field of AB genotypes in D with mean density $c\, da$, where we assume

$$\int p(t,a,x)\, da = 1.$$

Then the results of Theorems 1-2-3 carry over with

$$m(t,x) = c\int m(t,a,x)da = ce^{mt},$$

$$(6.1)\qquad v(t,x,y) = c\int v(t,a,x,y)\,da + m(t,x)m(t,y) + Q$$

$$= cVs^2 e^{mt}\int_0^t e^{ms}\, q(s,x,y)ds + c^2e^{2mt} + Q$$

where $q(t,x,y) = \int p(t,a,x)p(t,a,y)da$, and Q is asymptotically small for large t. The error Q depends on the variances and covariances of the initial random field; in particular it vanishes if the initial field is Poisson. Similarly

$$(6.2)\qquad v_u(t,x,y) = cVs^2 e^{mt}\int_0^t e^{ms}e^{-2us}\, q(s,x,y)ds$$

and $v_u(t,x,y)$ is a weak solution of

$$(6.3)\qquad \frac{\partial}{\partial t}v_u = (L_x^* + L_y^* + 2m - 2u)v_u + Vs^2\delta(x-y).$$

The corresponding equation in the classical (i.e., not a branching process approximation) case for the correlation coefficient of local gene frequencies (Fleming and Su (1973)) is

$$(6.4)\qquad \frac{\partial}{\partial t}H = (L_x^* + L_y^* - 4u)H + (C_1 - C_2H)\,\delta(x-y).$$

Now, assume for simplicity $p(t,a,x) = p(t,x,a)$. Then $q(t,x,y) = p(2t,x,y)$, and the asymptotic values of v and v_u involve potentials of $p(t,x,y)$. (If $p(t,a,x) = h(t,x-a)$, then in general $q(t,x,y)$ is the transition density of the symmetrized random motion corresponding to $p(t,a,x)$.) In particular, if $m = 0$, as t goes to infinity,

$$(6.5)\qquad v(t,x,y) = \text{Prob}(\,N_x(t) = 1,\ N_y(t) = 1)/dxdy$$

$$= c\ \text{Prob}(N_x(t) = 1 \,|\, N_y(t) = 1)/dx \longrightarrow \infty$$

whenever the process in D corresponding to $p(t,a,x)$ is recurrent in D.

In particular (6.5) holds for Brownian motion in R^d for $d = 1,2$ but not for $d = 3$. It is difficult to interpret (6.5), but it seems to be related to the misbehavior of critical branching processes. There is no difficulty if m is negative or $d \geqslant 3$. There is never any trouble with v_u for $m \leqslant 0$; indeed if $m = 0$ $(s = 1)$

$$v_u(t,x,y) \longrightarrow v_u(x,y) = Vc\, g(2u,x,y)$$

where $g(2u,x,y) = \int_0^{\infty} e^{-2us} q(s,x,y)\,ds.$

Similarly one can define $(x \neq y)$

$$(6.6) \qquad C(t,a,x,y) = \text{Prob}\left[\text{Individuals at } x,y \text{ are i.b.d.} \;\middle|\; \text{There are individuals at } x \text{ and } y\right]$$

$$= v_u(t,a,x,y)/v(t,a,x,y)$$

and

$$C(x,y) = \lim C(t,x,y), \qquad C(t,x,y) = v_u(t,x,y)/v(t,x,y).$$

For $m = 0$ in one and two dimensions, $C(x,y) = 0$ due to the catastrophe (6.5). For m negative, $C(x,y)$ is a ratio of exponentials in odd dimension and of Hankel functions in even dimensions. For $m = 0$ in three dimensions, $C(x,y)$ makes sense and is

$$C(x,y) = \frac{V\, e^{-|x-y|\sqrt{2u}}}{V + 4\pi c|x-y|}.$$

See Sawyer (1974) for more detail on these results.

REFERENCES

J. Crow and M. Kimura (1970), An introduction to population genetics theory, Harper and Row, New York.

D. Dawson (1972), Stochastic evolution equations, Math. Biosciences 15, 287-316, Appendix I.

D. Dawson (1974), Stochastic evolution equations and related measure processes, manuscript.

W. Fleming and C. Su (1973-74), One dimensional migration models in population genetics theory, ms., and Some one dimensional migration models in population genetics theory, Theor. Popn. Biol. 5, June 1974.

W. Fleming (1974a), Distributed parameter stochastic systems in population biology, manuscript.

N. Ikeda, M. Nagasawa, and S. Watanabe (1968-69), Branching Markov processes I,II,III, J. Math. Kyoto 8, 233-278, 365-410, and 9, 95-160.

M. Kimura and T. Ohta (1971), Theoretical aspects of population genetics, Monographs in Popn. Biol. No. 4, Princeton Univ. Press.

G. Malécot (1948), The mathematics of heredity, W. H. Freeman and Co., San Francisco, English translation 1969.

P. Moran, The statistical processes of evolutionary theory, Clarendon Press, Oxford, 1962.

T. Nagylaki (1974), The decay of genetic variability in geographically structured populations, Proc. NAC USA 71, 2932-2936.

S. Sawyer (1970), A formula for semi-groups, with an application to branching diffusion processes, Trans. AMS 152, 1-38.

S. Sawyer (1974), Branching diffusion processes in population genetics, manuscript.

T. Stewart (1973), The people of America, Charles Scribner, New York.

G. Weiss and M. Kimura (1964-65), The stepping stone model of population structure and the decrease of genetic correlation with distance, Genetics 49, 561-576, and A mathematical analysis of the stepping stone model of genetic correlation, Jour. Applied Prob. 2, 129-149.

Relativistic Brownian Motion

Jean-Pierre Caubet

After Planck's paper of 1900 about the black-body radiation law and Einstein's paper of 1905 about light quanta, L. de Broglie put forward his ideas about matter waves around 1923. Using Planck's relation $E = h\nu$, he proposed in particular the celebrated relation $\lambda = h/p$ which arises for instance from a proportion between the Poincaré-Cartan form and the wave phase. As this differential form also appears in Feynman's integral, no doubt (cf. the probabilistic approach to heat, i.e. infrared transfer, de Broglie's thermodynamics, Fenyes' and Nelson's papers,...) that it might be involved in diffusion processes. The reader will find in this paper how such an idea is at the root of the definition of the relativistic Brownian motion. In fact this form splits into two closed forms ω_1 and ω_2, and when they are exact (compare with the Hamilton-Jacobi theory), the wave function $\psi = \exp(R+iS)$ with $\omega_1 = d(\hbar R)$ and $\omega_2 = d(\hbar S)$ is a solution of the Klein-Gordon equation. Moreover, any relativistic spinorial equation governs relativistic Brownian motions with stationary spins. Performing a second quantization and assuming the nature of the coupling terms, we can then calculate the scattering amplitudes in the electromagnetic interactions, for instance as was done by Feynman around 1949.

1. Hamilton-Jacobi Theory

The aim of this section is to show how some standard facts of Hamilton-Jacobi theory can be derived from the factorization of observables (momentum, energy,...) through the local coordinates of the basic manifold (involving what will appear to be regression functions in the probabilistic construction of the relativistic brownian motion) when we are concerned with quadratic hamiltonians, avoiding so difficulties to transversality. In particular, the Jacobi function S comes by degenerescence from a regression function associated with the dynamic of this spacetime brownian motion.

The following result is enough to exemplify and illuminate the factorization effect.

Proposition 1. (Dynamic of a particle with electric charge). From the a priori given four-momentum relation

$$(p-KA)^2 + m_0^2 c^2 = 0 ,$$

p being the four-momentum, K the charge, A the four-vector electromagnetic potential, and m_0 the rest mass of the particle, we deduce (Legendre transformation) the relations

$$p\text{-}KA = m_0 v/\sqrt{1-\beta^2} = m_0 u = mv$$

in which $\beta^2 = v^2/c^2$. Then the equations of motion are

$$d\, mv_\lambda = K(\partial_\lambda A_\mu - \partial_\mu A_\lambda)\, dq^\mu = K\, H_{\lambda\mu}\, dq^\mu$$

in which q^μ is the (contravariant) four-vector, H the skew-symmetric electromagnetic tensor of order 2, and dq^μ the real four-displacement.

Proof. We only need to use the Hamilton equations.

Proposition 2. (Hamilton-Jacobi theory for a particle with electric charge). We now assume the factorization of p through q . Then the Poincaré-Cartan form $\theta = p\, dq$ is closed, and so, exact as well. In other words, there exists a Jacobi function $S : M \to R$ (i.e. defined on the Minkowski space M , and real valued) such that $dS = \theta$, this function being necessarily a solution of the Jacobi equation

$$(\nabla S - KA)^2 + m_0^2 c^2 = 0 ,$$

Proof. From the equations of motion we deduce that

$$dp_\lambda = K\partial_\lambda A_\mu \, dq^\mu ,$$

whence

$$dp_\lambda - (\partial_\lambda p_\mu)\, dq^\mu = -(\partial_\lambda mv_\mu)\, dq^\mu = 0 .$$

So the form θ is closed ; it is exact as well (Poincaré lemma), in other words the function S does exist, and as we have $p = \nabla S$, the four-momentum relation makes the function S satisfy the Jacobi equation.

2. Schrödinger Equation

We define and construct here the diffusion processes determined by solutions of the Schrödinger equation and begin by describing the notation used. Let $M = R^3$ be the 3-dimensional Euclidean space, (q^j) where $j = 1,2,3$ the coordinates of $q \in M$, X a diffusion process in M and $X_t = (X_t^j)$ where $j = 1,2,3$ the position of X at time t . We denote the left and right derivatives of X by

$$(Lf)(q) = \lim_{s \uparrow t} E\left[\frac{f(X_t)-f(X_s)}{t-s} \;\middle|\; X_t = q\right] = \partial_t f + \langle v_-, \nabla f\rangle - \nu\Delta f$$

and

$$(Rf)(q) = \lim_{u \downarrow t} E\left[\frac{f(X_u)-f(X_t)}{u-t} \;\middle|\; X_t = q\right] = \partial_t f + \langle v_+, \nabla f\rangle + \nu^*\Delta f$$

where $\langle \ , \ \rangle$ denotes the inner product, ∂_t , $\nabla =$ grad and $\Delta =$ div grad denote the partial derivative $\partial/\partial t$, the gradient and the Laplace operator acting on differentiable functions f on M , and $E[\,|\,]$ denotes the conditional expectation.

Proposition 2. (Continuity of the diffusion process). Under classical assumptions of regularity concerning v_+ , ν^* and the initial distribution of X , the operators $-L$ and R are adjoined to each other with the measure ρdq , in other words, whatever the functions with compact support $f, g : M \to R$ are, we have the following relation

$$(1) \qquad \int f\,(Lg)\,\rho\, dq + \int (Rf)\, g\,\rho\, dq = 0 ,$$

where $\rho : M \to R$ denotes the (density of the) distribution of the process X at time t , so that

$$\rho(q^j)\, dq^j = P\{X_t^j \in dq^j\}$$

where $dq^j \equiv dq$ denotes the Lebesgue measure on M. Moreover we have

$$\nu = \nu^* \qquad\qquad v_+ = v_- + 2\rho^{-1}\nabla(\nu\rho)$$

and the following Fokker-Planck relations hold

$$L^+\rho = \partial_t\rho + \operatorname{div}(\rho v_-) + \Delta\nu\rho = 0 \;, \quad -R^+\rho \equiv \partial_t\rho + \operatorname{div}(\rho v_+) - \Delta\nu\rho = 0 \;,$$

where L^+ and R^+ are respectively adjoined to L and R with the Lebesgue measure dq, whence we deduce, in using the notation $v = 2^{-1}(v_+ + v_-)$, the continuity equation

$$\partial_t\rho + \operatorname{div}(\rho v) = 0 \;.$$

Proof. Let $t_j = a + j\,[(b-a)/n] = a + j\epsilon$ $(0 \leqslant j \leqslant n)$ be a partition of the time interval (a, b). After having suitably conditioned the right part of the following elementary relation

$$E\,[f(X_b)\, g(X_{t_{n-1}}) - f(X_{t_1})\, g(X_a)] =$$

$$\epsilon \sum_{j=1}^{n-1} E\left[\frac{f(X_{t_{j+1}}) - f(X_{t_j})}{\epsilon}\;\frac{g(X_{t_j}) + g(X_{t_{j-1}})}{2}\right] +$$

$$\epsilon \sum_{j=1}^{n-1} E\left[\frac{f(X_{t_{j+1}}) + f(X_{t_j})}{2}\;\frac{g(X_{t_j}) - g(X_{t_{j-1}})}{\epsilon}\right]$$

whith respect to P_{t_j} and F_{t_j}, where $P_t = \sigma(X_s \,;\, s \leqslant t)$ and $F_t = \sigma(X_s \,;\, s \geqslant t)$ denote the past and future σ-algebras of events generated by the process X at time t, we get when $\epsilon \downarrow 0$ (i.e. when $n \uparrow \infty$)

$$E\,[f(X_b)\, g(X_b) - f(X_a)\, g(X_a)] =$$

$$\int_a^b dt\; E\,[(f(Lg) + (Rf)g)\,(X_t)] \;,$$

and so, letting $b-a \downarrow 0$, we deduce that

$$\frac{d}{dt}\, E\,[f(X_t)\, g(X_t)] = E\,[(f(Lg) + (Rf)g)\,(X_t)] \;.$$

Integrating the two parts of this last relation over all the real line R (with respect to the Lebesgue measure dt), we get the relation 1. This relation leads to $L = -\rho^{-1} R^{+}\rho$. And as we can directly calculate R^{+}, we get

$$R^{+}(\rho g) = -\partial_t(\rho g) - \langle v_+, \nabla \rho g\rangle - \rho g \operatorname{div} v_+ + \Delta \nu^{*} \rho g$$

So, using the second Fokker-Planck equation 3, we now have the formula

$$\rho^{-1} R^{+}(\rho g) = -\partial_t g - \langle v_+, \nabla g\rangle + \nu^{*}\Delta g + 2\rho^{-1}\langle \nabla \nu^{*}\rho, \nabla g\rangle,$$

which we identify to Lg, so as to get the relations 2. Finally, to establish the Fokker-Planck equations directly (since we have just used the second one), we have only to observe that the relation 1 can also be written in the two following ways

$$\int f\,[\rho Lg + R^{+}(\rho g)]\,dq = 0 \quad \text{and} \quad \int f[\rho Rg + L^{+}(\rho g)]\,dq = 0 ,$$

whence, since the function f is arbitrary, we deduce the relations $\rho L + R^{+}\rho = 0$ and $\rho R + L^{+}\rho = 0$. But if the function g equals one over the support of the function f, these two relations reduce to Fokker-Planck equations $R^{+}\rho = 0$ and $L^{+}\rho = 0$.

Proposition 3. (Wave function). We put $\nu = \hbar/2M$ and so we define the mass M, necessarily strictly positive, of the particle. The form

$$\omega_1 = \langle M\delta v , dq\rangle - E_{osm}\,dt ,$$

where $\delta v = 2^{-1}(v_+ - v_-)$ and $E_{osm} = \nu \operatorname{div} Mv + \langle v, M\delta v\rangle$, is closed by the continuity of the process, and so it is exact as well. More precisely, if we define the scalar function $R : M \rightarrow R$ by the relation $\nu\rho = \nu_0 \exp(2R)$, ν and ν_0 having the same dimension, we have $\omega_1 = d\hbar R$.

More-over, under the assumption that the form

$$\omega_2 = \langle Mv, dq\rangle - Edt ,$$

where E denotes the (not yet defined) energy of the particle, is closed and so exact as well, there exists a function $S : M \rightarrow R$ (we make it scalar by introducing the multiplicative constant $\hbar$), extension of the classical Jacobi function, such that $\omega_2 = d(\hbar S)$.

Putting $\psi = \exp(R+iS)$, we then have

$$\rho = \frac{\nu_0}{\nu} |\psi|^2 .$$

Proof. The closeness of ω_1 is due to the continuity of the process, more precisely it derives from the second relation (2) and from the continuity equation (3) in the preceding Proposition 3. The remnant part of the proof is straightforward.

We now define the energy E so as to associate the Schrödinger equation with the diffusion process.

Proposition 4. (Schrödinger equation). We now assume that the diffusion coefficient $\nu = \hbar/2M$ is a constant. We also assume that the equations of motion are

$$\operatorname{curl} Mv = 0 \qquad\qquad L\, Mv_+ + R\, Mv_- = 0 .$$

Then the form

$$\omega_2 = \langle Mv, dq \rangle - E dt ,$$

where $E = 2^{-1} M [v^2 - (\delta v)^2] - \nu \operatorname{div} M \delta v$, is closed ; and the wave function $\psi = \exp(R + iS)$ is a solution of the Schrödinger equation

$$i \frac{\partial \psi}{\partial t} = - \nu \Delta \psi .$$

Proof.Thereis only a straightforward verification to perform. The closeness of ω_2, expressed in terms of observables, gives in particular the equation of motion

$$2^{-1}(L\, Mv_+ + R\, Mv_-) = \partial_t Mv + \langle v, \nabla Mv \rangle - \langle \delta v, \nabla M \delta v \rangle - \nu \Delta M \delta v = 0 .$$

Moreover, separating the Schrödinger equation into real and imaginary parts, we let the continuity equation and the equation of motion both appear.

Throughout this text, we call "translated form" and "translated wave equation" any form or wave equation associated with a particle interacting with some field.

Proposition 5. (Translated Schrödinger equation). We still assume that the diffusion coefficient is a constant, and that we have the energy E given as above in Proposition 5. But we now give us real-valued potentials A, V defined on M .

The form ω_1 is obviously unchanged, and we now assume the closeness of the translated form

$$\omega_2 = \langle Mv + A, dq \rangle - (E+V)\, dt .$$

Then the corresponding wave function $\psi = \exp(R+iS)$ is a solution of the translated Schrödinger equation

$$(i\hbar \partial_t - V)\psi = \frac{1}{2M}\left(\frac{\hbar}{i}\nabla - A\right)^2 \psi .$$

Proof. The verification is straightforward. In the particular case of an electron interacting with the electromagnetic field, we then have the Maxwell equations

$$\mathcal{E} + \partial_t A = -\nabla V \qquad \mathcal{H} = \operatorname{curl} A$$

so the form ω_2 is closed by the following equations of motion

$$\operatorname{curl}(Mv+A) = 0 \qquad 2^{-1}(LMv_+ + RMv_-) = \mathcal{E} + v \times \mathcal{H},$$

where $\mathcal{E} + v \times \mathcal{H}$ stands for the classical Laplace force. We emphasize that in such a force, no osmotic term (i.e. no term involving δv) appears, a fact broken down by the curvature, if any, of the basic manifold M.

Before going on for taming the curvature velleity, let us remark that the Proposition 3 is general. It works in particular for the classical brownian motion, and then we have

$$-\omega_1 = \omega_2 = \langle \frac{q}{2t}, dq \rangle - \frac{q^2 - t}{4t^2}\, dt .$$

And it as well works for the relativistic brownian motion, involving what is usually called the fifth dimension. Moreover, let us remark that the time parameter t ranges a necessarily open interval T so as to let the proof of the Proposition 3 be performed : we can observe that the form ω_1, associated with the classical brownian motion is not defined at $t = 0$. The last remark to do, but not the least, is that the wave function gives us the initial data and the derivative of the process. In other words, the wave function completely determines the diffusion process, its construction appears to be a classical one.

3. Pauli Equation

By induction, we are straight forward led to give us a d-dimensional Riemannian manifold M, the first Betti number of which being zero. Define (q^j) where $j = 1,\dots, d$ to be the local coordinates of $q \in M$. The diffusion process ranging M has the position $X = (X_t^j)$ at time $t \in T$, where T denotes an open interval of the real line. The left and right derivatives of X are defined as above and we still have

$$(Lf)(q) = \partial_t f + \langle v_-, \nabla f \rangle - \nu \Delta f$$

$$(Rf)(q) = \partial_t f + \langle v_+, \nabla f \rangle + \nu^* \Delta f$$

but now with $\langle\ ,\ \rangle$ the inner product on tangent vectors, ∇ = grad and Δ = div grad the gradient and the Laplace-Beltrami operator acting on (indefinitely) differentiable functions f on M.

Proposition 6. (Schrödinger equation, Riemannian case). We still assume that the diffusion coefficient $\nu = \hbar/2M$ is a constant.

Then the form

$$\omega_1 = \langle M \delta v, dq \rangle - E_{osm}\, dt ,$$

where $\delta v = 2^{-1} (v_+ - v_-)$ and $E_{osm} = \nu \operatorname{div} Mv + \langle v, M\delta v \rangle$, is closed by the continuity of the process, and so there exists a scalar function $R : M \to R$ such that $\omega_1 = d(\hbar R)$.

Moreover, we assume the closeness of the form

$$\omega_2 = \langle Mv, dq \rangle - Edt$$

where $E = 2^{-1} M \left[v^2 - (\delta v)^2\right] - \nu \operatorname{div} M\delta v$, and we write it $\omega_2 = d(\hbar S)$. So the wave function $\psi = \exp (R+iS)$ is a solution of the Schrödinger equation

$$i \frac{\partial \psi}{\partial t} = - \nu \Delta \psi .$$

Proof. Only standard Riemannian technique is needed.

The osmotic terms involved by centrifugal forces are tamed, since the assumed closeness of the form ω_2 gives us the equation of motion. And it appears that we could no longer consider the extension of the classical $f = m\gamma$ law as the fundamental effect in Physics, since we cannot guess what are the osmotic terms of the force as they are much too complicated, when the closeness of the form ω_2 appears to be a much simpler phenomenon : we can already infer that the closeness of the form ω_2 is more fundamental, and that the composite expression of the energy E only comes from the degenerescence of the relativistic case which we expect to be simpler.

To exemplify the curvature effect, we now define the quantized diffusion process on the Riemannian manifold $M = R^3 \times SU(2)$, whith $SU(2)$ the group of unitary 2x2 matrices of determinant +1 which is the universal covering space of the special orthogonal group $SO(3)$. Translating the corresponding Schrödinger equation, we will then have the Pauli equation at hand.

Proposition 7 (Schrödinger equation on $R^3 \times SU(2)$). Given $M = R^3 \times SU(2)$, we assume the diffusion coefficients $\nu_1 = \hbar/2M$ and $\nu_2 = \hbar/2I$, with M the mass and I the moment of inertia, isotropic in both the spaces R^3 and $SU(2)$, but not necessarily equal to each other.

Then the wave function $\psi = \exp(R+iS)$ associated with the quantized free diffusion on $R^3 \times SU(2)$ is a solution of the Schrödinger equation

$$i\frac{\partial\psi}{\partial t} = -(\nu_1\Delta_1 + \nu_2\Delta_2)\psi$$

with Δ_1 and Δ_2 the Laplace-Beltrami operators on R^3 and $SU(2)$ respectively.

Assuming now that the particle is interacting with some field, in the translated form ω_2 extra terms appear and we have

$$\omega_2 = \langle Mv + A, dq\rangle_1 + \langle Is + B, dq\rangle_2 - (E + V)\,dt$$

where $\langle\ ,\ \rangle_1$ and $\langle\ ,\ \rangle_2$ denote the inner products in R^3 and $SU(2)$, and where $s(=$ spin$)$ stands for v on $SU(2)$. The translated Schrödinger equation is

$$(i\hbar\partial_t - V)\psi = \frac{1}{2M}\left(\frac{\hbar}{i}\nabla_1 - A\right)^2\psi + \frac{1}{2I}\left(\frac{\hbar}{i}\nabla_2 - B\right)^2\psi\ ,$$

with ∇_1 and ∇_2 the gradients on R^3 and $SU(2)$, but we want to define quantized diffusion processes on R^3 with a stationary spin.

Introducing the Euler angles, the generic matrix in $SU(2)$ can be written

$$\begin{pmatrix} \cos\frac{\theta}{2}\exp\left[\frac{i}{2}(\varphi+\chi)\right] & i\sin\frac{\theta}{2}\exp\left[\frac{i}{2}(\varphi-\chi)\right] \\ i\sin\frac{\theta}{2}\exp\left[-\frac{i}{2}(\varphi-\chi)\right] & \cos\frac{\theta}{2}\exp\left[-\frac{i}{2}(\varphi+\chi)\right] \end{pmatrix}$$

Let us still introduce the three following operators

$$M_x = \cos\varphi \frac{\partial}{\partial\theta} + \frac{\sin\varphi}{\sin\theta}\left(\frac{\partial}{\partial\chi} - \cos\theta\frac{\partial}{\partial\varphi}\right)$$

$$M_y = \sin\varphi \frac{\partial}{\partial\theta} - \frac{\cos\varphi}{\sin\theta}\left(\frac{\partial}{\partial\chi} - \cos\theta\frac{\partial}{\partial\varphi}\right) \quad \text{and} \quad M_z = \frac{\partial}{\partial\varphi} .$$

Each entry of the generic matrix above is an eigenvalue of the operator $i\,M_z$ corresponding to the eigenvalue $\pm\frac{1}{2}$, and it is also an eigenvalue of the Laplace-Beltrami operator $\Delta_2 = M_x^2 + M_y^2 + M_z^2$ corresponding to the eigenvalue $-2^{-1}(2^{-1}+1)$. So we now have the complete proof of the following result at hand.

Proposition 8. (Pauli equation). Let the assumptions be the same as in Proposition 7, then any wave function

$$\psi = F_+ \otimes \psi_+ + F_- \otimes \psi_- ,$$

where $\binom{\psi_+}{\psi_-}$ denotes any column of the generic matrix above and where F_+, F_- are complex-valued functions only defined on R^3, which is a solution of the Schrödinger equation

$$(ih\partial_t - V)\psi = \frac{1}{2M}\left(\frac{\hbar}{i}\nabla_1 - A\right)^2\psi + \frac{1}{2I}\left(\frac{\hbar}{i}\nabla_2 - B\right)^2\psi ,$$

is such that, writing now

$$\psi = \binom{F_+}{F_-} ,$$

this column vector is by itself a solution of the spinorial equation

$$(ih\partial_t - V)\psi = \frac{1}{2M}\left(\frac{\hbar}{i}\nabla_1 - A\right)^2\psi$$

$$+ \frac{1}{2I}\left[\frac{3}{4}\hbar^2 - 2\langle B,\sigma\rangle_2 - \frac{\hbar}{i}\,\mathrm{div}_2 B + B^2\right]\psi$$

with $\langle B,\sigma\rangle_2 = B_x\sigma_x + B_y\sigma_y + B_z\sigma_z$, where B_x, B_y and B_z denote the components of B with respect to the frame (M_x, M_y, M_z) — we identify any vector in each tangent vector space of $SU(2)$ with the Lie derivative that this direction defines — , and where σ_x, σ_y, σ_z denote the Pauli matrices

$$\sigma_x = \begin{pmatrix} 0 & 1 \\ 1 & 0 \end{pmatrix} \qquad \sigma_y = \begin{pmatrix} 0 & -i \\ i & 0 \end{pmatrix} \qquad \sigma_z = \begin{pmatrix} 1 & 0 \\ 0 & -1 \end{pmatrix} .$$

4. Relativistic Brownian Motion

Up to now, we have considered that the extension of the classical $f = m\gamma$ law was the right one, as far as the force f did not involve any osmotic term. We still maintain this attitude, so as to define the relativistic Brownian motion in the simplest case, and we will induce the general case in which centrifugal forces are involved by assuming the closeness of the well identified form ω_2. But then, we will analyse the converse attitude : Is the closeness of ω_2 enough to define the relativistic Brownian motion ? At first, the answer seems to be no, but things become clearer when we involve stability assumptions, following a de Broglie's idea.

We start off with the notation used. Let $\mathbb{M}$ be the Minkowski space, $(q^o = ict, q^j)$ where $j = 1,2,3$ the coordinates of $q \in \mathbb{M}$, X a diffusion process in $\mathbb{M}$ and $X_t = (X_t^o = ic\ T_t, X_t^j)$ where $j = 1,2,3$ the position of X at time t. Do always take in mind that the Minkowski time (in ict) and the parameter time (in X_t) are not the same, although we use the same letter for both of them as no confusion cannot really happen : factorizing the process completely through the Minkowski coordinates, the parameter time will disappear. We should only have trouble with the fifth dimension in the forms ω_1 and ω_2, by applying Proposition 4, but we are very little concerned with.

We denote the left and right derivatives of X by

$$(Lf)(q) = \lim_{s\uparrow t} E\left[\frac{f(X_t)-f(X_s)}{t-s} \mid X_t = q\right] = \langle v_-, \nabla f\rangle - \nu\Box f$$

and

$$(Rf)(q) = \lim_{u\downarrow t} \left[\frac{f(X_u)-f(X_t)}{u-t} \mid X_t = q\right] = \langle v_+, \nabla f\rangle + \nu^*\Box f$$

where $\langle\ ,\ \rangle$ denotes the inner product, ∇ = grad and $\Box$ = div grad denote the four-dimensional gradient and Laplacian operating on (indefinitely) differentiable functions f on $\mathbb{M}$. As usual, $E[\,|\,]$ still denotes the conditional expectation.

For any particle, we assume that

$$v_o = ic$$

with v_o the speed of the particle in the Minkowskian time direction, so that $\langle v, v\rangle = v_j v^j - c^2$. That is the assumption which will make the parameter time vanished in the forms ω_1 and ω_2 (and so in the wave equation also). But we could make it reappeared by adding to both these forms a fifth term, exactly as it was done in Proposition 3. Our last assumption is the following stationariness property :

$$P\left\{X_t^j \in dq^j \mid T_t = s\right\} = P\left\{X_s^j \in dq^j \mid T_s = s\right\} ,$$

whatever the parameter time t may be.

Proposition 9. (Continuity of the spacetime diffusion). Under classical assumptions of regularity concerning v_+, ν^* and the initial distribution of X, the operators $-L$ and R are adjoined to each other with respect to the measure ρdq, in other words, whatever the functions with compact support $f, g : M \to R$ are, we have the following relation

(1) $$\int f(Lg)\rho dq + \int (Rf)g\rho dq = 0 ,$$

where $\rho : M \to R$ denotes the density of the conditional distribution of the process at time t such that

$$\rho(ict, q^j)\, dq^j = P\left\{X_t^j \in dq^j \mid T_t = t\right\},$$

with $dq^j = dq$ the Lebesgue measure on M. Moreover we have

(2) $$\nu = \nu^* \qquad v_+ = v_- - 2\rho^{-1}\nabla(\nu\rho) ,$$

and also the following Fokker-Planck equations

(3) $$L^+\rho \equiv \operatorname{div}(\rho v_-) + \Box\nu\rho = 0 , \quad -R^+\rho \equiv \operatorname{div}(\rho v_+) - \Box\nu\rho = 0 ,$$

where L^+ and R^+ are respectively adjoined to L and R with respect to the Lebesgue measure dq, whence we deduce, still using the notation $v = 2^{-1}(v_+ + v_-)$, the continuity equation

$$\operatorname{div}(\rho v) = 0 .$$

Proof. Let $t_j = a + j\, n^{-1}(b-a) = a + j\epsilon\ (0 \leq j \leq n)$. After having conditioned the two parts of the following elementary relation

$$E[f(X_b)\, g(X_{t_{n-1}}) - f(X_{t_1})\, g(X_a)] =$$

$$\sum_{j=1}^{n-1} \int P\left\{T_{t_j} \in ds\right\} E\left[\frac{f(X_{t_{j+1}}) - f(X_{t_j})}{\epsilon}\ \frac{g(X_{t_j}) + g(X_{t_{j-1}})}{2} \mid T_{t_j} = s\right]$$

$$+ \sum_{j=1}^{n-1} \int P\left\{T_{t_j} \in ds\right\} E\left[\frac{f(X_{t_{j+1}}) + f(X_{t_j})}{2}\ \frac{g(X_{t_j}) - g(X_{t_{j-1}})}{\epsilon} \mid T_{t_j} = s\right]$$

with respect to the event $\bigcap_{a\leq t\leq b} \{a \leq T_t \leq b\}$, we still condition both the terms of its second part suitably with respect to the σ-algebras P_{t_j} and F_{t_j}, where $P_t = \sigma(X_s\ ;\ s \leq t)$ and $F_t = \sigma(X_u\ ;\ u \geq t)$. Thus this relation, in the limit as $\epsilon \downarrow 0$ (i.e. as $n \uparrow \infty$), reduces to

$$E\,[f(X_b)\,g(X_b) - f(X_a)\,g(X_a) \mid a \leq T_t \leq b] =$$

$$\int_a^b dt\; E\,[(f(Lg) + (Rf)\,g)\,(X_t) \mid T_t = t]$$

which, in the limit as $b-a \downarrow 0$, gives us

$$\frac{d}{dt} E\,[f(X_t)\,g(X_t) \mid T_t = t] = E[(f(Lg) + (Rf)\,g)(X_t) \mid T_t = t]\,.$$

By integrating the two parts of this last relation over all the real line (with respect to the Lebesgue measure dt), we get the relation 1 . This relation gives us $L = -\rho^{-1} R^+\rho$, and as we can directly calculate R^+, so as to get

$$R^+(\rho g) = -\langle \rho v_+, \nabla g\rangle - g\,\mathrm{div}(\rho v_+) + \Box(\nu \rho g)\,,$$

by using the second Fokker-Planck equation 3, we now have the formula

$$\rho^{-1} R^+(\rho g) = -\langle v_+, \nabla g\rangle + \nu\Box g + 2\rho^{-1}\langle \nabla \nu\rho, \nabla g\rangle$$

which we identify to Lg so that we get the relations 2. Let us now go back to the Fokker-Planck equations. We observe that the relation 1 can also be written in the two following ways

$$\int f\,[\rho Lg + R^+(\rho g)]\,dq = 0 \qquad \text{and} \qquad \int f\,[\rho Rg + L^+(\rho g)]\,dq = 0\,,$$

which give us the Fokker-Planck equations $R^+\rho = 0$ and $L^+\rho = 0$ if the function g equals one over the support of the (arbitrary but fixed) function f .

The following proposition gives us the correct probabilistic interpretation of the wave function in the relativistic case.

Proposition 10. (Wave function). We write $\nu = \hbar/2M$ and so we introduce the mass M, necessarily strictly positive, of the particle. The form

$$\omega_1 = \langle M\delta v,\ dq\rangle\,,$$

with $\delta v = 2^{-1}(v_+ - v_-)$, is closed by the continuity of the process, and exact. If we define the scalar function $R : \mathbb{M} \rightarrow R$ by the relation $\nu\rho = \nu_o \exp(2R)$, with ν_o a constant having the dimension of ν, we have $\omega_1 = d(\hbar R)$.

On the other hand, under the assumption that the form

$$\omega_2 = \langle Mv, dq \rangle$$

is closed, and then exact, there exists a function $S : \mathbb{M} \rightarrow R$ (we make it scalar by introducing the Planck's constant $\hbar$) such that $\omega_2 = d(\hbar S)$.

Putting $\psi = \exp(R+iS)$, we then have

$$\rho = \frac{\nu_o}{\nu} |\psi|^2 .$$

Proof. The exactness of the form ω_1 is only due to the relation between v_+ and v_- in Proposition 9, we have neither to use the continuity equation to check the closeness first nor to apply the Poincaré lemma so as to assert the existence of the function R. But should we want to introduce the fifth dimension, we would have to proceed so, with

$$E_{osm} = \nu(\operatorname{div} Mv + \partial_t Mv) + \langle v, M\delta v \rangle .$$

We go on to define the relativistic Brownian motion.

Proposition 11. (Klein-Gordon equation). We now assume that the equation of motion is

$$2^{-1} (L\, Mv_+ + R\, Mv_-) = 0 .$$

Then the diffusion coefficient $\nu = h/2M$ is necessarily defined by

$$M = M_o / \sqrt{1-\beta^2} \qquad \text{with } M_o\, c = \sqrt{m_o^2\, c^2 - \hbar^2\, (\Box e^R)/e^R}$$

in which m_o denotes a constant and $1-\beta^2 = -c^{-2}\langle v, v \rangle$, and we have the relation

$$E = \langle v, Mv \rangle_N - L \qquad \text{with } E = Mc^2,\ L = -M_o c^2 \sqrt{1-\beta^2} ,$$

in which $\langle\ ,\ \rangle_N$ denotes the three-dimensional scalar product in the real vector subspace N of $\mathbb{M}$.

Moreover, if we still assume that $\operatorname{curl}_N Mv = 0$, then the form

$$\omega_2 = \langle Mv, dq \rangle$$

is closed and exact, and the wave function $\psi = \exp(R+iS)$ is a solution of the Klein-Gordon equation

$$\Box \psi = \frac{m_o^2 c^2}{\hbar^2} \psi .$$

Proof. We start off with the following identity

$$2^{-1}(L M_+ + RMv_-) = \langle v, \nabla Mv \rangle - \langle v, \nabla M \delta v \rangle - \nu \Box M \delta v$$

which gives us the four equations of motion

$$\frac{d}{dt} [Mv_\lambda] = \frac{\hbar^2}{2M} \partial_\lambda \left(\frac{\Box e^R}{e^R}\right) .$$

So we have the relation

$$\langle Mv, Mv \rangle - \hbar^2 \frac{\Box e^R}{e^R} = - m_o^2 c^2 = \text{const.}$$

which gives us M_o. Moreover, the relation $M = M_o / \sqrt{1-\beta^2}$ is identical to the relation $E = \langle v, Mv \rangle_N - L$. The remnant part of the proof is straight forward.

As the diffusion coefficient is now identified we can translate the form ω_2.

Proposition 12. (Translated Klein-Gordon equation). We assume the diffusion coefficient $\nu = \hbar/2M$ defined as in Proposition 11, and the closeness of the form

$$\omega_2 = \langle Mv + A, dq \rangle ,$$

in which A denotes the four-vector potential.

Then the corresponding wave function $\psi = \exp(R+iS)$ is a solution of the translated Klein-Gordon equation

$$-\left(\frac{\hbar}{i} \partial_\mu - A\right)\left(\frac{\hbar}{i} \partial^\mu - A\right) \psi = m_o^2 c^2 \psi .$$

And at this stage, we can construct the process : from stationariness assumption, it follows that we have

$$\lim_{h \downarrow 0} E \left[\frac{f(X_{.+h}) - f(X_.)}{h} \mid X_. = q \right] = \langle v_+, \nabla f \rangle + \nu \Box f ,$$

whatever the dotted parameter time may be, all we have to do now is to give us the initial data about T_t and the wave function which then determines the process thoroughly, the construction still appears to be a classical one.

Conversely, let us assume the closeness of the form

$$\omega_2 = \langle Mv, dq \rangle .$$

We still write the diffusion coefficient $\nu = \hbar/2M$ with $M = M_o/\sqrt{1-\beta^2}$. Then we get the following equations of motion

$$\frac{d}{dt}\,[Mv_\lambda] = -\,c\sqrt{1-\beta^2}\;\partial_\lambda(M_o c) \;,$$

but we a priori know nothing about M_o , we should be wrong in thinking that M_o is now determined. For instance, both the equations of motion

$$L\,Mv_+ + R\,Mv_- = 0 \qquad \text{and} \qquad L\,Mv_- + R\,Mv_+ = 0$$

are compatible with the equations of motion just above, and they respectively give us $M_o^2\,c^2$ equal to

$$m_o^2\,c^2 - \hbar^2\,\frac{\Box e^R}{e^R} \qquad \text{and} \qquad m_o^2\,c^2 + \hbar^2\,\frac{\Box e^R}{e^R} \;.$$

We assume that we only have to choose between both of them, in regard that they are the simplest ones. Then we cannot accept the second law if we realize that, in an equilibrium, the wave function is monochromatic : taking the second equation of motion as the right one, such an equilibrium would be unstable and the diffusion coefficient would be minimal, both things which sound unnaturally. So, assuming the closeness of ω_2, whether we deduce the law $LMv_+ + RMv_- = 0$ by degenerescence and then induce it up to the Minkowski space, or we infer the law from stability considerations, we are led to the same definition of the relativistic Brownian motion. And we keep the Lorentz invariance safe.

So, we will now take the closeness of the form ω_2 as a rule. If we observe that the same property holds for the three-dimensional classical Brownian motion and its spatial form ω_2 (i.e. for its form ω_2 minus the term Edt), a fact due to the relations

$$v_+ = 0 \quad , \qquad \int_C \langle Mv_- \,,\, dq \rangle = \int_C \langle Mv_+ \,,\, dq \rangle$$

whatever the closed curve C may be, we are led to the conclusion that the relativistic Brownian motion we have just defined is the natural extension of the classical Brownian motion compatible with the Minkowskian structure.

5. Dirac Equation

We can now proceed to define a diffusion X ranging the Minkowski space $\mathbb{M}$ by a solution of the Dirac equation. Starting off with the notation used, define $(q^o = ict, q^j)$ with $j = 1,2,3$ to be the coordinates on $\mathbb{M}$, and with $j = 1,2,3,\ldots 9$ the local coordinates on the product $M = \mathbb{M} \times SL(2,C)$, where $SL(2,C)$ denotes the group of complex 2×2 matrices of determinant $+1$ (which is the universal covering group of the proper Lorentz group). On $SL(2,C)$ we will define the Riemannian metric by taking $ds^2 = 2(dt)^2$ trace $(\dot{S}\,\dot{S}^*)$ with S^* and S the inverse (with respect to the group structure) and the derivative (with respect to the parameter time t) of $S \in SL(2,C)$. Define $X_t = (X_t^o = ic\, T_t, X_t^j)$ with $j = 1,\ldots,9$ the position of X at time t. The Minkowski time in ict and the parameter time in X_t are not the same.

We denote the left and right derivatives of X by

$$(Lf)(q) = \langle v_-, \nabla f \rangle - \nu \Box f \qquad (Rf)(q) = \langle v_+, \nabla f \rangle + \nu \Box f$$

where $\langle\ ,\ \rangle$ denotes the inner product on M, ∇ = grad and $\Box$ = div grad denote the gradient and Laplace-Beltrami operator acting on (indefinitely) differentiable functions f on $M = \mathbb{M} \times SL(2,C)$.

For any particle, we still assume that

$$v_o = ic$$

with v_o the time component of the speed v of the particle, and we assume the following stationariness property

$$P\left\{ X_t^j \in dq^j \mid T_t = s \right\} = P\left\{ X_s^j \in dq^j \mid T_t = s \right\},$$

whatever the parameter time t may be.

Proposition 13. (Klein-Gordon equation on $\mathbb{M} \times SL(2,C)$). We now assume that the form

$$\omega_2 = \langle Mv, dq \rangle$$

is closed, so that $\omega_2 = d(\hbar S)$ for some scalar functions, and we induce that $\nu = \hbar/2M$ with

$$M = M_o/\sqrt{1-\beta^2}, \qquad M_o\, c = \sqrt{m_o^2\, c^2 - \hbar^2(\Box e^R)/e^R},$$

m_o a constant and $1-\beta^2 = -c^{-2}\langle v, v\rangle$. We then have the relation

$$E = \langle v, Mv\rangle_{N\times SL(2,C)} - L \qquad \text{with} \quad E = Mc^2 \ , \ L = -M_o\, c^2\sqrt{1-\beta^2} \ ,$$

in which $\langle \ , \ \rangle_{N\times SL(2,C)}$ denotes the scalar product on $N \times SL(2,C)$ (N still denotes the real vector subspace of $\mathbb{M}$) .

Moreover, the wave function $\psi = \exp\,(R+iS)$ is a solution of the Klein-Gordon equation

$$\Box\,\psi = \frac{m_o^2\, c^2}{\hbar^2}\,\psi \ .$$

We can now proceed to pass to the Dirac equation. The Lie algebra of $SL(2,C)$ is generated by the following matrices

$$A_{01} = \frac{1}{2}\,(\sigma_3 \otimes \sigma_1) \qquad\qquad A_{23} = \frac{1}{2i}\,(1 \otimes \sigma_1)$$

$$A_{02} = \frac{1}{2}\,(\sigma_3 \otimes \sigma_2) \qquad\qquad A_{31} = \frac{1}{2i}\,(1 \otimes \sigma_2)$$

$$A_{03} = \frac{1}{2}\,(\sigma_3 \otimes \sigma_3) \qquad\qquad A_{12} = \frac{1}{2i}\,(1 \otimes \sigma_3)$$

where $\sigma_1 = \sigma_x$, $\sigma_2 = \sigma_y$ and $\sigma_3 = \sigma_z$ denote the three Pauli matrices. And if we define the matrices μ_j, τ_j with $j = 1,2,3$ by the relations

$$A_{oj} = \frac{1}{i}\,(\tau_j - \mu_j) \qquad\qquad A_{ki} = (\tau_j + \mu_j)$$

$SL(2,C)$ appears to be the direct product of two groups G_r $(r=1,2)$ having the same Lie algebra as $SU(2)$, for we now have $[\mu_i, \tau_j] = 0$ and

$$[\mu_1, \mu_2] = \mu_3, \ \ldots \qquad\qquad [\tau_1, \tau_2] = \tau_3, \ldots$$

cyclically. Let $\binom{e_+}{e_-}$ denote a fixed column of the generic matrix of $SU(2)$, and let us search for wave functions ψ , solutions of the Klein-Gordon equation just above, but with a stationary spin, that is to say with

$$\psi = \sum_{\pm\pm} \psi_{\pm\pm}\; e_{\pm} \otimes e_{\pm} \ .$$

Then, using the following result about the Dirac equation, we will construct four diffusion processes ranging the Minkowski space and conditionned by what is the corresponding stationnary spin.

Proposition 14. (Dirac equation). Any solution of the Dirac equation

$$\left(\frac{1}{c}\partial_t - \sum_{j=1}^{3} \alpha_j \partial_j + \frac{i}{\hbar}\mathbf{m}_o c \alpha_o\right)\psi = 0$$

with $\alpha_o = \sigma_1 \otimes 1$, $\alpha_j = \sigma_3 \otimes \sigma_j$ and $\psi = (\psi_{\pm\pm})$, is such that the corresponding wave function

$$\psi = \sum_{\pm\pm} \psi_{\pm\pm} \; e_{\pm} \otimes e_{\pm}$$

is a solution of the Klein-Gordon equation on $M \times SL(2,C)$ if the following relation holds

$$\mathbf{m}_o^2 = m_o^2 + \frac{3}{2}\frac{\hbar^2}{c^2}$$

Proof. Define Δ_{SL} to be the Laplace-Beltrami operator on $SL(2,C)$. We then have

$$\Delta_{SL} \; e_{\pm} \otimes e_{\pm} = -\frac{3}{2} e_{\pm} \otimes e_{\pm} ,$$

and the verification is now straight forward.

We can also translate the Dirac equation, but there is no connection with the translated Klein-Gordon equation on $M \times SL(2,C)$, unless we neglect the quadratic potential terms. In fact, the translated Dirac equation is associated with the wave function corresponding to the energetic invariant relation

$$\langle p-A, p-A\rangle_M + \langle p, p\rangle_{SL} - 2\langle p, H\rangle_{SL} + m_o^2 c^2 = 0 ,$$

in which $\langle \, , \, \rangle_M$ and $\langle \, , \, \rangle_{SL}$ denote the inner products in M and $SL(2,C)$, and H denotes the skew-symmetric electromagnetic tensor of order 2. In other words, the translated form ω_2 is unchanged, but the equation of motion is different, and the diffusion is still constructed via the wave function exactly as above. Then squared mass M_o^2 is translated, and that is just why the equation of motion is changed, a fact which also works for second order non-linear wave equations.

6. Relativistic Spinorial Equations

In the last section, we have associated solutions of the Dirac equation with the relativistic Brownian motion on $M \times SL(2,C)$, the spin of which being stationary and equal to 1/2. In the same way, we can associate the linear relativistic spinorial equations with Brownian motions on the groups $M \times (SL(2,C))^n$ $(n \geq 1)$. For instance, if we consider the group $M \times (SL(2,C))^2$, we get the following result :

Proposition 15. (Brownian motion with stationary spin ≤ 1). Any solution of the compatible $(i = 1,2)$ system

$$\left(\frac{1}{c}\partial_t - \sum_{j=1}^{3} \alpha_j^i \partial_j + \frac{i}{\hbar}\mathfrak{m}_o c\,\alpha_o^i\right)\psi = 0 ,$$

where $\alpha_j^1 = \alpha_j \otimes 1$, $\alpha_j^2 = 1 \otimes \alpha_j$ with $\alpha_j = \sigma_3 \otimes \sigma_j$ $(j = 1,2,3)$, $\alpha_j = \sigma_1 \otimes 1$ $(j=0)$, and where $\psi = (\psi_{\underline{++++}})$ denotes a sixteen-vector, is such that the corresponding wave function

$$\psi = \sum_{\underline{++++}} \psi_{\underline{++++}}\, e_{\underline{+}} \otimes e_{\underline{+}} \otimes e_{\underline{+}} \otimes e_{\underline{+}}$$

is a solution of the Klein-Gordon equation on $M \times (SL(2,C))^2$ if the following relation holds

$$\mathfrak{m}_o^2 = m_o^2 + 2 \cdot \frac{3}{2}\,\frac{\hbar^2}{c^2}$$

The column vector $\psi = (\psi_{\underline{++++}})$ has sixteen components, ten of which define the Brownian motion on M conditional on the spin 1 $(= 2^{-1} + 2^{-1})$, the six others define the Brownian motion on M conditional on the spin 0 $(= 2^{-1} + 2^{-1})$. It is a well known fact that the ten components associated with the spin 1 are governed by Maxwellian equations which, in the limit as $\mathfrak{m}_o \to 0$, become the classical Maxwell equations.

7. Antiparticles

Up to now, the diffusion processes that we have constructed in the preceding sections are what we will call "particles", they are in particular defined by the condition $v_o = ic$. If instead of this condition, we assume that $v_o = -ic$, we will speak of "antiparticles" running backward in time.

We still define the left and right derivatives of such processes $X_t = (X_t^o = ic\, T_t, X_t^j)$ by

$$Lf = \langle v_-, \nabla f \rangle - \nu \Box f \qquad\qquad Rf = \langle v_+, \nabla f \rangle + \nu^* \Box f ,$$

and we will also write $\nu = \hbar/2M$. Be well aware of the fact that, for both particles and antiparticles, mass, energy and probability are constantly positive, though the time component $M\, v_o$ of the momentum is (if we neglect the complex factor i) positive for particles, negative for antiparticles.

To construct the antiparticle process, we will assume the following stationariness property :

$$P\left\{ X_t^j \in dq^j \mid T_t = -s \right\} = P\left\{ X_s^j \in dq^j \mid T_s = -s \right\}$$

whatever the parameter time t may be, and we will determine the (left or right) derivative of the process (i.e. the mass M and the speed v_+ or v_-) via the wave function according to the following result, given for an antiparticle ranging the Minkowski space $\mathbb{M}$

Proposition 16. (Antiparticles). The operators $-L$ and R are adjoined to each other with respect to ρ times the Lebesgue measure dq , in other words, whatever the functions with compact support $f, g : \mathbb{M} \to R$ may be, we have the following relation

$$(1) \qquad \int f(Lg)\, \rho\, dq + \int (Rf)\, g \rho\, dq = 0$$

where $\rho : \mathbb{M} \to R$ denotes the density of the process at time t conditional on $T_t = -t$, so that

$$\rho\,(-ict, q^j)\, dq^j = P\left\{ X_t^j \in dq^j \mid T_t = -t \right\} ,$$

with dq^j the Lebesgue measure on N . We still have the Fokker-Planck equations $L^+\rho = 0$ and $R^+\rho = 0$, the relations

$$\nu = \nu^* \qquad\qquad v_+ = v_- + 2\, \rho^{-1} \nabla\, (\nu \rho) ,$$

and the equation of continuity $\operatorname{div}(\rho\, v) = 0$.

Moreover, assuming the closeness of the form

$$\omega_2 = \langle Mv\, , dq \rangle$$

we induce the relations

$$M_o c = \sqrt{m_o c^2 - \hbar^2 (\Box e^R)/e^R} \qquad \rho = \frac{M}{m_o} e^{2R}$$

with m_o the (constant) rest mass of the antiparticle, and the wave function is a solution of the Klein-Gordon equation

$$\Box \psi = \frac{m_o^2 c^2}{\hbar^2} \psi .$$

Proof. It is quite similar to the particle case, we start off with the same elementary relation, but we consider it conditional on $\bigcap_{a \leq t \leq b} \left\{ -b \leq T_t \leq -a \right\}$, so as to get

$$\frac{d}{dt} E \left[f(X_t) g(X_t) \mid T_t = -t \right] = E \left[((Rf)g + f(Lg)) (X_t) \mid T_t = -t \right]$$

Let us now emphasize the difference between particles and antiparticles, when operators are involved. We can first observe that, for both particles and antiparticles, we have

$$(\hbar/2im_o) \left(\bar{\psi} \nabla \psi - \psi \nabla \bar{\psi} \right) = \rho v ,$$

and the time component of this four-vector is $\rho v_o = \pm ic$. In other words, the (positive) distribution density ρ arises from this relation, but multiplied by $-ic$ in the antiparticle case. Moreover, define the mean value of an observable to be its integral with respect to exp (2R) times the Lebesgue measure dq^j in the real subspace N . As we have $\rho v = m_o^{-1} \exp(2R)\, Mv$, the following result where we have systematically associated the operator $i\hbar \partial_t$ with the energy appears at once :

Proposition 18. (Mean value of the four-momentum). Let ψ be the wave function associated with a particle or an antiparticle. Then the mean value of Mv_j and $E = Mc^2$ are, respectively, given by

$$\int_N \bar{\psi} \frac{\hbar}{i} \partial_j \psi \, dq^j \quad \text{and} \quad \int_N \bar{\psi}\, i\hbar\, \partial_t \psi \, dq^j$$

in the particle case, and in the antiparticle case by

$$\int_N \psi \frac{-\hbar}{i} \partial_j \bar{\psi} \, dq^j \quad \text{and} \quad \int_N \psi\, i\hbar\, \partial_t \bar{\psi} \, dq^j$$

So the Dirac equation, when it is associated with an antiparticle, will now be written as follows :

$$(i\hbar\,\partial_t + \sum_{j=1}^{3} c\,\alpha_j \frac{-\hbar}{i}\,\partial_j + m_o c^2 \alpha_o)\,\bar{\psi} = 0 .$$

And we will write in the same way any relativistic spinorial equation if it is associated with an antiparticle, so as to let the energy appear.

8. Second Quantization

We start off with the spinorial equation $i\hbar\,\partial_t \psi = H\psi$ associated with a free Brownian motion on $M \times SL(2,C)^m$ and define (ψ_i) to be a complete orthonormal set of states, i.e. of eigenfunctions ψ_i of the Hamiltonian operator H . Then we construct the multiparticle space $H^+ = \bigoplus_{n\geq 0} H_n^+$ with $H_o^+ = C$ and H_n^+ the n-particle (vector) space spanned by the n-particle states

$$\psi = \psi_{i_1} \circ \ldots \circ \psi_{i_n} = \frac{1}{\sqrt{n!}} \sum_{\sigma} \epsilon^{\sigma} \psi_{\sigma(i_1)} \otimes \ldots \otimes \psi_{\sigma(i_n)} ,$$

where $\epsilon = +1$ (Bose case) or -1 (Fermi case), and where σ runs through all permutations of n objects. The inner product of the two n-particle states φ and ψ is given by

$$\langle \varphi , \psi \rangle = \sum_{\sigma} \epsilon^{\sigma} \langle \varphi_{i_1}, \psi_{\sigma(i_1)} \rangle \cdots \langle \varphi_{i_n}, \psi_{\sigma(i_n)} \rangle .$$

Moreover, H_n and H_m are orthogonal when $n \neq m$, in other words, we define states of different number of particles to be orthogonal to each other. We can see that for a complete orthomornal set of states in the multiparticle space H we can take

$$| \, n_1 \, n_2 \ldots \rangle = \frac{\psi_{i_1} \circ \ldots \circ \psi_{i_n}}{\sqrt{n_1! n_2! \ldots}}$$

where n_i is the number of times that i occurs in the sequence $i_1, \ldots, i_n$. Of course it goes without saying that $n_i = 0$ or 1 in the Fermi case. Next we define the creation and annihilation operators through

$$a^+(\psi)(\psi_1 \circ \ldots \circ \psi_n) = \psi \circ \psi_1 \circ \ldots \circ \psi_n$$

$$a(\psi)(\psi_1 \circ \ldots \circ \psi_n) = \sum_{k=1}^{n} \epsilon^{k-1} \langle \psi , \psi_k \rangle (\psi_1 \circ \ldots \circ \hat{\psi}_k \circ \ldots \circ \psi_n) ,$$

in which $\hat{\psi}_k$ stands for no ψ_k .

Proposition 17. (Free Brownian motion on $\bigcup_{n\geq 1} (M \times SL(2,C)^m)^n$).

The following wave equation

$$\sum_{k=1}^{n} \psi_1 \circ \ldots \circ \left(\square - \frac{m_o^2 c^2}{\hbar^2}\right) \psi_k \circ \ldots \circ \psi_n = 0 ,$$

in which $\square$ denotes the Laplace-Beltrami operator on $M \times SL(2,C)^m$, is associated with any free particle (or antiparticle) ranging the manifold $M \times SL(2,C)^m$. More generally, with any one particle operator A is associated the (derivation) operator A such that

$$A(\psi_1 \circ \ldots \circ \psi_n) = \sum_{k=1}^{n} \psi_1 \circ \ldots \circ A\psi_k \circ \ldots \circ \psi_n .$$

In particular, to the identity operator in the one particle space corresponds the number of particles operator $N = \sum_i a^+(\psi_i)\, a(\psi_i)$.

The proof is based on linearity (limits, derivations, conditionnings, ..) and stochastic independence of particles belonging to a same multiparticle.

Let us now assume that the one particle is interacting with some field, so that the spinorial equation is now translated, and we have $i\hbar\partial_t \psi = (H+V)\psi$. Let us keep the states defined just as above (no perturbation case) et let us develop the generic state ψ of the multiparticle space :

$$\psi = \sum_{n_1 n_2} R(n_1 n_2 \ldots t) \mid n_1 n_2 \ldots >$$

If there were no perturbation, the distribution function $R(n_1 n_2 \ldots t)$ would not depend on the time, but since we now assume that the particle is interacting with some field, the operators $i\hbar\partial_t$ and $H+V$ which act on the one particle space, do also act on the multiparticle space in the way described in Proposition 18. So we now have the relation (we use the bracket notation)

$$i\hbar\partial_t R(n_1 n_2 \ldots t) = \sum_{m_1 m_2 \ldots} R(m_1 m_2 \ldots t) < n_1 n_2 \ldots |V| m_1 m_2 \ldots > ,$$

in the right part of which the scalar products are either equal to zero or given through one of the two following relations

$$\langle \ldots \circ \psi_i^{(n_i)} \circ \ldots , \ldots \circ \psi_i^{(n_i-1)} \circ V \psi_i \ldots \rangle = (n_i!)\, V_{ii}$$

$$(n_j+1) \langle \ldots \circ \psi_i^{(n_i)} \circ \psi_j^{(n_j)} \circ \ldots , \ldots \circ \psi_i^{(n_i-1)} \circ \psi_j^{(n_j)} \circ V \psi_j \ldots \rangle = (n_i!)(n_j+1)!\, V_{ij} ,$$

with $\psi_i^{(n_i)} = \psi_i \circ \ldots \circ \psi_i (n_i$ times) and $V_{ij} = \langle \psi_i, V\psi_j \rangle = \int_N \bar{\psi}_i V \psi_j \, dq^j$.

So the relation just above can also be written

$$i\hbar\, \partial_t R(n_1 n_2 \ldots t) = \sum_i V_{ii}\, n_i R(\ldots n_i \ldots t) + \sum_{ij} V_{ij} \sqrt{n_i(n_j+1)}\, R(\ldots n_i-1, n_j+1 \ldots t)$$

That is the equation of evolution in the multiparticle space, it can be written in the following convenient form

$$i\hbar\, \partial_t R = H R \qquad \text{with } H = \sum_{ij} c_i c_j^+ \int_N \bar{\psi}_i V \psi_j \, dq^j ,$$

where c_i and c_j^+ denote the annihilation and creation operators defined just below.

Proposition 19. (Mean value in the multiparticle space). In the particle case, the mean value of an observable being given by

$$\langle \psi, A\psi \rangle = \int_N \bar{\psi} A \psi \, dq^j ,$$

then the corresponding mean value of this observable in the multiparticle space H^+ is given by

$$\langle \psi, A\psi \rangle_{H^+} = \sum_{n_i n_j} \bar{R}\,(n_1 n_2 \ldots) A_{ij}\, c_i c_j^+ R\,(n_1 n_2 \ldots),$$

where $A_{ij} = \langle \psi_i, A\psi_j \rangle = \int_N \bar{\psi}_i A \psi_j \, dq^j$, $c_i = \sqrt{n_i}\, \exp(-\partial/\partial n_i)$ (annihilation) and $c_j^+ = \exp(\partial/\partial n_j) \sqrt{n_j}$ (creation).

Proof. Going back to the relations above, we first have

$$\langle \psi, A\psi \rangle_H = \sum_{n_1 n_2 \ldots} \bar{R}\,(n_1 n_2 \ldots) \langle n_1 n_2 \ldots \mid A \mid m_1 m_2 \ldots \rangle R\,(m_1 m_2 \ldots)$$

$$= \sum_{n_i} \bar{R}\,(\ldots n_i \ldots)\, A_{ii}\, n_i\, R(\ldots n_i \ldots)$$

$$+ \sum_{n_i n_j} \bar{R}\,(\ldots n_i \ldots)\, A_{ij} \sqrt{n_i(n_j+1)}\, R\,(\ldots n_i-1, n_j+1 \ldots)$$

Then we use the annihilation and creation operators c_i and c_i^+, which satisfy the following relations

$$c_i\, R(\ldots n_i \ldots) = \sqrt{n_i}\, R(\ldots n_i - 1 \ldots) \qquad c_i^+\, R(\ldots n_i \ldots) = \sqrt{n_i+1}\, R(\ldots n_i + 1 \ldots)$$

The result of our efforts is that we are now led to the quantum field theory formalism. We begin by formulating the preceding results in the antiparticle case. We first give us a spinorial equation $i\hbar \partial_t \bar{\psi} = H \bar{\psi}$, next we construct the multiantiparticle space $H^- = \bigoplus_{n \geq 0} H_n^-$ with $H_o^- = C$ and H_n^- the n-antiparticle space spanned by the n-antiparticle states

$$\psi = \psi_{j_1} \circ \ldots \circ \psi_{j_n} = \frac{1}{\sqrt{n!}} \sum_{\sigma} \epsilon^{\sigma} \psi_{\sigma(j_1)} \otimes \ldots \otimes \psi_{\sigma(j_n)} ,$$

in which $\epsilon = +1$ (Bose case) or -1 (Fermi case). If now the one-antiparticle is interacting with some field, we have the translated spinorial equation $i\hbar \partial_t \bar{\psi} = (H+V)\bar{\psi}$, and the generic state

$$\psi = \sum_{n_1 n_2 \ldots} R(n_1 n_2 \ldots t) \mid n_1 n_2 \ldots >$$

of the multiantiparticle space satisfies the following equation

$$i\hbar\, \partial_t \bar{R} = H \bar{R} \qquad \text{with } H = \sum_{ij} c_i\, c_j^+ \int_N \psi_i\, V \bar{\psi}_j\, dq^j .$$

Moreover, if the mean value of an observable is given by

$$< \psi , A\psi > = \int_N \psi\, A\, \bar{\psi}\, dq^j ,$$

then the corresponding mean value of this observable in the multiantiparticle space $\bar{H}$ is given by

$$< \psi , A\psi >_{H^-} = \sum_{n_i n_j} R(n_1 n_2 \ldots)\, A_{ij}\, c_i\, c_j^+\, \bar{R}\, (n_1 n_2 \ldots)$$

where $A_{ij} = \int_N \psi_i\, A\, \bar{\psi}_j\, dq^j$.

We are now ready to introduce the operators $a_i^+ = c_i$ and $a_i = c_j^+$, in order to express that annihilating an antiparticle is equivalent to creating energy, and vice versa. Instead of writing the generic state of the multiparticle and antiparticle spaces as we have done up to now, we write $\psi = \sum c_i^+ \psi_i$ (particle case) and $\psi = \sum a_j \psi_j$ (antiparticle case). Inthe particle case, the dependance on the time will be said regular ($\sim \exp(-i\epsilon t)$ with $\epsilon \geqslant 0$), and in the antiparticle case it will be said irregular ($\sim \exp(i\epsilon t)$ with $\epsilon \geqslant 0$), for in that case the energy seems to be negative, though only the time component of the fourmomentum is negative. So we can at last consider a mixture ψ of particles and antiparticles, and its conjugate ψ^+

$$\psi = \sum (c_i^+ \psi_i + a_j \psi_j) \qquad \psi^+ = \sum (c_i \bar{\psi} + a_j^+ \bar{\psi}_j)$$

which appear as Fourier expansions : any calculus performed on particles and antiparticles may then be translated in terms of these Fourier series, so as to give us the quantum field theory formalism. As an example, the study of electromagnetic interactions between electrons and photons is exactly the same in both these formalisms, the differences are only formal :

The translated Dirac equation involves the coupling term

$$e \langle \bar{\psi} \gamma^\mu \psi , A_\mu \rangle$$

where $\langle \, , \, \rangle$ denotes the inner product in the Minkowski space, and where γ^μ are the Heisenberg matrices. So the equations of evolution in the multiparticle space involve the terms

$$e \, c_i \, c_j^+ \int_N (\bar{\psi}_i \gamma^\mu \psi_j A_\mu) \, dq^j$$

and the second order term in the Dyson-Feynman expansion is

$$\frac{e^2}{2!} \iint T \, [(\bar{\psi}_i \gamma^\mu \psi_j A_\mu)(q) \, c_i \, c_j^+ (\bar{\psi}_k \gamma^\nu \psi_l A_\nu)(q') \, c_k \, c_l^+] \, dq \, dq'$$

with T the time-ordered operator. The part of this term involving the operator $c_4 \, c_3 \, c_2^+ \, c_1^+$ is this operator times

$$e^2 \, [\bar{\psi}_4 \gamma^\mu \psi_2 \cdot \bar{\psi}_3' \gamma^\nu \psi_1' - \bar{\psi}_4 \gamma^\mu \psi_1 \cdot \bar{\psi}_3' \gamma^\nu \psi_2'] \, T \, [A_\mu \, A_\nu'] \, dq \, dq'$$

with $\psi = \psi(q)$, $\psi' = \psi(q')$, $A_\mu = A_\mu(q)$ and $A_\nu' = A_\nu(q')$. Just at this point we have made the junction with the classical formalism, and we now go on as usual with monochromatic wave functions $\psi_i(q) = u_i(q) \exp(i \langle p_i , q \rangle)$ so as to get by Fourier transform the relations $p_3 - p_2 = p_1 - p_4$ and $p_3 - p_1 = p_2 - p_4$ between four-momenta corresponding

respectively, to the Feynman diagrams

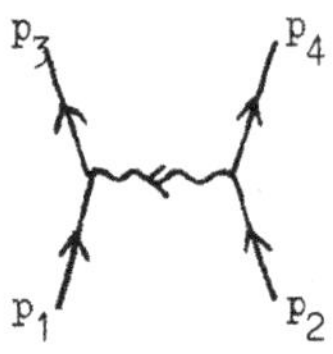

and

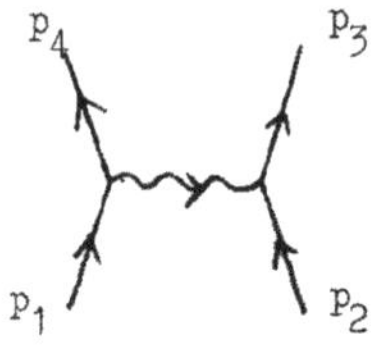

and the amplitudes of both these scatterings.

Bibliography

BROGLIE, L. de

[1] "Recherches sur la Théorie des Quanta". Thesis, Paris, 1924 ; Ann. Phys., 3, 22-128, 1925.

[2] "Mécanique Ondulatoire du Photon et Théorie Quantique des Champs". Gauthiers-Villars, Paris, 1957.

[3] "Thermodynamique de la Particule Isolée ou Thermodynamique Cachée des Particules". Gauthier-Villars, Paris, 1966.

CARTAN, E.

[1] "Leçons sur les Invariants Intégraux". Hermann, Paris, 1958.

CAUBET, J.P.

[1] Dynamique de la Diffusion Spatiale, Spin et Equation de Pauli, C.R. Acad. Sc. Paris, 274, 1502-1505, 1972.

[2] Equation de Dirac et Dynamique du Mouvement Brownien Spatio-temporel, C.R. Acad. Sc. Paris, 276, 887-890, 1973.

[3] Sur la Fonction d'Onde en Théorie Quantique Relativiste, C.R. Acad. Sc. Paris, 277, 1199-1202, 1973.

[4] Sur la Seconde Quantification, C.R. Acad. Sc. Paris, 278, 1271-1273, 1974.

[5] Sur les Antiparticules, C.R. Acad. Sc. Paris, 279, 247-250, 1974.

FENYES, I.

[1] Eine Wahrscheinlichkeitstheoretische Begründung und Interpretation der Quantenmechanik, Zeitschrift für Physik 132, 81-106 (1952).

FEYNMAN, R.P.

[1] The Theory of Positrons, Physical Review, 76, 749-759, 1949.

[2] Space-Time Approach to Quantum Electrodynamics, Physical Review, 76, 769-789, 1949.

FEYNMAN, R.P., and HIBBS, A.R.

[1] "Quantum Mechanics and Path Integrals". McGraw-Hill, New York, 1965.

GANGOLLI, R.

[1] On the Construction of Certain Diffusions on a Differentiable Manifold, Z. Wahrscheinlichkeitsh 2, 209-419 (1964.

GIHMAN, J.J., and SKOROHOD, A.V.

[1] "Stochastic Differential Equations". Springer, Berlin, 1972.

ITO, K., and McKEAN, H.P., Jr.

[1] "Diffusion Processes and Their Sample Paths." Springer, Berlin, 1965.

McKEAN, H.P., Jr.

[1] "Stochastic Integrals". Academic Press, New York, 1969.

NELSON, E.

[1] "Dynamical Theories of Brownian Motion". Princeton University Press, Princeton, 1967.

RAIS, M.

[1] Les Solutions Invariantes de l'Equation des Ondes, Comptes Rendus Acad. Sc. Paris, 259, 2169-2170, 1964.

STROOCK, D.W., and VARADHAN, S.R.S.

[1] Diffusion Processes with Continuous Coefficients, 1, 2, Com. on Pure and Appl. Math. 22, 345-400, 1969.

Asymptotics and Limit Theorems for the Linearized Boltzmann Equation[1]

Richard S. Ellis[2]

We consider the initial value problem for the linearized Boltzmann equation

(1) $$\frac{\partial p}{\partial t} + \xi \cdot \text{grad } p = \frac{1}{\epsilon} Qp, \quad \lim_{t \downarrow 0} p = f,$$

where the initial data $f = f(x,\xi)$ and the solution $p = p_\epsilon(t,x,\xi)$, $t > 0$, $x \in \mathbb{R}^3, \xi = (\xi_1, \xi_2, \xi_3) \in \mathbb{R}^3$. Q is the linearized collision operator corresponding to a spherically symmetric intermolecular potential, and $\epsilon > 0$ is a parameter which represents the mean free path. Corresponding to the conservation of number, momentum, and energy in an individual collision, Q has a five-dimensional nullspace spanned by 1, ξ_i $(i = 1,2,3)$, and $|\xi|^2$.

We are interested in the asymptotic behavior of the solution of (1) as $\epsilon \downarrow 0$. This is formally treated at the physical level of rigor by the Chapman-Enskog-Hilbert procedure [9; pp. 254-262]. Given a nice f, we define

1. This is a report on joint work with Mark A. Pinsky.

2. Supported in part by National Science Foundation Grant GP-28576

$$f_0(x) \equiv \langle f(x,\cdot),\ 1\rangle;\quad f_i(x) \equiv \langle f(x,\cdot),\ \xi_i\rangle,\ i = 1,2,3;$$

$$f_4(x) \equiv \langle f(x,\cdot),\ \frac{|\xi|^2 - 3}{\sqrt{6}}\rangle; \tag{2}$$

$$n_0(t,x) \equiv \langle p_\epsilon(t,x,\cdot),\ 1\rangle;\quad n_i(t,x) \equiv \langle p_\epsilon(t,x,\cdot),\ \xi_i\rangle,\ i = 1,2,3;$$

$$n_4(t,x) \equiv \langle p_\epsilon(t,x,\cdot),\ \frac{|\xi|^2 - 3}{\sqrt{6}}\rangle,$$

where p_ϵ is the solution of (1) with initial data f and where $\langle\ ,\ \rangle$ denotes the inner product in the Hilbert space $\mathscr{H}_0 = L^2(\mathbb{R}^3;\ (2\pi)^{-3/2}\exp(-\frac{1}{2}|\xi|^2)d\xi)$. Provided f satisfies certain algebraic compatibility conditions (the Hilbert relations), one can show that up to an error $O(\epsilon)$ the n_i satisfy the following system of partial differential equations (linear Euler equations):

$$\frac{\partial n_0}{\partial t} = -\operatorname{div}\vec{n},$$

$$\frac{\partial \vec{n}}{\partial t} = -\operatorname{grad} n_0 - \sqrt{\frac{2}{3}}\operatorname{grad} n_4, \tag{4}$$

$$\frac{\partial n_4}{\partial t} = -\sqrt{\frac{2}{3}}\operatorname{div}\vec{n},$$

$$n_i(0^+,x) = f_i(x),$$

where $\vec{n} = (n_1,n_2,n_3)$. Further, if one writes $p_\epsilon(t/\epsilon,x,\cdot)$ for $p_\epsilon(t,x,\cdot)$ in (2), then up to an error $O(\epsilon)$ the resulting $n_i(\frac{t}{\epsilon},x)$ agree with the solutions (evaluated at time t/ϵ) of the following system of partial differential equations (linear Navier-Stokes equations):

$$\frac{\partial n_0}{\partial t} = - \operatorname{div} \vec{n},$$

(4) $$\frac{\partial \vec{n}}{\partial t} = - \operatorname{grad} n_0 - \sqrt{\frac{2}{3}} \operatorname{grad} n_4 + \epsilon\eta[\Delta\vec{n} + \frac{1}{3} \operatorname{grad} \operatorname{div} \vec{n}],$$

$$\frac{\partial n_4}{\partial t} = - \sqrt{\frac{2}{3}} \operatorname{div} \vec{n} + \epsilon\lambda\Delta n_4,$$

$$n_i(0^+,x) = f_i(x).$$

In (4), $\epsilon > 0$, $n_i = n_i^\epsilon(t,x)$, $i = 0,\ldots,4$, and $\eta > 0$ and $\lambda > 0$ are physical constants. Notice that the Euler equations are obtained from (4) by setting $\epsilon = 0$.

In a series of basic papers [7,8,10], Grad has studied the existence of the solution of (1) and has sought to make precise the formal results obtained by the Chapman-Enskog-Hilbert procedure. He begins with the decomposition, valid for a class of so-called cut-off hard potentials:

(5) $$Q = \nu - K.$$

Here ν is the operator of multiplication by the collision frequency $\nu(\xi)$, a strictly positive function of $|\xi|$, and K is a compact operator on $\mathscr{H}_0$. Using (5), Grad writes (1) as an integral equation and then derives a priori estimates in the Hilbert space $\mathscr{H} \equiv L^2(R^6;\ (2\pi)^{-3/2} \exp(-\frac{1}{2}|\xi|^2)d\xi dx)$. Concerning the asymptotic behavior of p_ϵ, let A and B denote, respectively, the first-order and second-order spatial partial derivatives in (4). Defining

$$p_\varepsilon \equiv T_\varepsilon(t)f, \ \exp(t(A + \varepsilon B))f = n_0^\varepsilon + \sum_{i=1}^{3} n_i^\varepsilon \xi_i + n_4^\varepsilon \frac{|\xi|^2 - 3}{\sqrt{6}},$$

(n_i^ε, $i = 0,\dots,4$, solve the Navier-Stokes system (4)), Grad proves the following asymptotic results, which are valid for any $f \in \mathcal{H}$ satisfying a mild growth and smoothness condition:

$$\text{(6)} \qquad T_\varepsilon(t)f = \exp(tA)f + O(\varepsilon), \quad \text{as} \quad \varepsilon \downarrow 0,$$

$$\text{(7)} \qquad T_\varepsilon(t/\varepsilon)f = \exp(\tfrac{t}{\varepsilon}(A + \varepsilon B))f + O(\varepsilon), \quad \text{as} \quad \varepsilon \downarrow 0.$$

In physical terms, (6) describes the non-viscous fluid approximation at a fixed time $t > 0$; (7) describes the viscous effects when $t \to \infty$. Our aim is to show that (7) is only one of a large variety of possible refinements of (6). This is accomplished by the following two results.

BOLTZMANN LIMIT THEOREM. *Let $f \in \mathcal{H}$ be sufficiently regular. Then*

$$\text{(8)} \qquad \exp(-tA/\varepsilon)T_\varepsilon(t/\varepsilon)f = \overline{N}(t)f + O(\varepsilon), \quad \text{as} \quad \varepsilon \downarrow 0,$$

where $\overline{N}(t)$ is a contraction semigroup on H whose generator is given by the differential equations

$$\frac{\partial n_0}{\partial t} = \left(\frac{9}{25}\lambda + \frac{2}{5}\eta\right)\Delta n_0 + \sqrt{\frac{2}{3}}\left(-\frac{6}{25}\lambda + \frac{2}{5}\eta\right)\Delta n_4,$$

$$\text{(9)} \qquad \frac{\partial \vec{n}}{\partial t} = \eta\Delta\vec{n} + \left(\frac{\lambda}{5} - \frac{\eta}{3}\right)\text{grad div } \vec{n},$$

$$\frac{\partial n_4}{\partial t} = \sqrt{\frac{2}{3}}\left(-\frac{6}{25}\lambda + \frac{2}{5}\eta\right)\Delta n_0 + \left(\frac{11}{25}\lambda + \frac{4}{15}\eta\right)\Delta n_4,$$

$n_i(0^+,x) = f_i(x)$;

i.e., $\overline{N}(t)f = n_0 + \sum_1^3 n_i\xi_i + n_4 \frac{|\xi|^2 - 3}{\sqrt{6}}$. *The semigroup* $\{\overline{N}(t), t \geq 0\}$ *commutes with the Euler semigroup* $\{\exp(tA), t \geq 0\}$.

In order to make the connection with (7), we also need the following.

NAVIER-STOKES LIMIT THEOREM. Let $f \in \mathcal{H}$ *be sufficiently regular. Then*

$$(10) \quad \exp(-tA/\epsilon)\exp(\tfrac{t}{\epsilon}(A + \epsilon B))f = \overline{N}(t)f + O(\epsilon), \quad \text{as} \quad \epsilon \downarrow 0.$$

The proof of (10) proceeds by means of Fourier transformation from the following purely algebraic result.

MATRIX LIMIT THEOREM.[3] *Let* A *be a skew-symmetric* $m \times m$ *matrix and* B *a real, symmetric, negative semidefinite* $m \times m$ *matrix. Let "exp" denote matrix exponentiation. Then*

$$\exp(-tA/\epsilon)\exp(\tfrac{t}{\epsilon}(A + \epsilon B)) = \exp(t\pi_A B) + O(\epsilon), \quad \text{as} \quad \epsilon \downarrow 0,$$

when $\pi_A B$ *is the orthogonal projection, in the Euclidean space of* $m \times m$ *matrices, of* B *onto the linear subspace of matrices which commute with* A.

In particular, we show that the generator of $\overline{N}(t)$ is $\pi_A B$, the projection of the Navier-Stokes operator B upon the set of operators which commute with the Euler operator A (B and A do *not* commute).

---

3 T. Kato (preliminary report) has generalized this result to the case of operators on a Banach space.

Using (10) and the commutativity of $\pi_A B$ and A, we have

(11) $$T_\epsilon(t/\epsilon)f = \exp(\tfrac{t}{\epsilon}(A + \epsilon\pi_A B))f + O(\epsilon), \quad \text{as} \quad \epsilon\downarrow 0.$$

This is the simplest of an infinite number of alternatives to (7). Indeed, we can show the existence of infinitely many solution operators $\exp(t\tilde{B})$ of parabolic systems of partial differential equations (one needs $\pi_A\tilde{B} = \pi_A B$) such that (11) remains true when $\pi_A B$ is replaced by $\tilde{B}$. This illustrates the asymptotic non-uniqueness of the Navier-Stokes equations. Further, since one of these $\tilde{B}$ is the Navier-Stokes operator B, we obtain an independent derivation of Grad's result (7) with, it turns out, weaker assumptions on f.

The proof of (8) depends on a careful spectral analysis of the operator $Q - i(\nu \cdot \xi)$, where $\nu\in R^3$ is a parameter. We prove the existence and differentiability, for $|\nu|$ sufficiently small, of the hydrodynamical eigenvalues and eigenfunctions $\{\alpha^{(j)}(\nu),\ e^{(j)}(\nu);\ j = 1,\dots,5\}$, which satisfy $\alpha^{(j)}(0) = 0$, $e^{(j)}(0)\in$ span $\{1,\xi_1,\xi_2,\xi_3,|\xi|^2\}$. We then prove a contour integral representation

(12) $$\exp[t(Q - i(\nu\cdot\xi)]f =$$

$$= \sum_{j=1}^{5} e^{t\alpha^{(j)}(\nu)} \langle f, e^{(j)}(-\nu)\rangle\, e^{(j)}(\nu) +$$

$$(2\pi i)^{-1} \int_C e^{t\alpha} R(\alpha,\nu) \frac{(Q - i(\nu\cdot\xi))^2 f}{\alpha^2}\, d\alpha ,$$

where C is a vertical contour in the half plane $\mathrm{Re}\alpha < 0$ and $R(\alpha,\nu) \equiv (Q - i(\nu\cdot\xi) - \alpha)^{-1}$. The first term of (12) corresponds to the Hilbert solution and gives the connection with the Euler and Navier-Stokes equations. The second term is negligible when Q/ε is written for Q and $\varepsilon \downarrow 0$. In case $\nu(\xi) \sim |\xi|^{\alpha}$ as $|\xi| \to \infty$ $(\alpha > 0)$, the contour integral may be replaced by $\int_C e^{t\alpha} R(\alpha,\gamma)\, f d\alpha$, where the contour C is such that $\mathrm{Re}\alpha \to -\infty$ $\mathrm{Im}\alpha \to \pm\infty$. The existence of the eigenvalues $\alpha^{(j)}(\nu)$ follows by applying the implicit function theorem to the exact hydrodynamical dispersion law. Previously, exact dispersion laws were obtained [15] only for hard sphere potentials, i.e., $\nu(\xi) \sim |\xi|$ as $|\xi| \to \infty$. In this case, the $\alpha^{(j)}(\nu)$ are analytic functions and can also be obtained from Rellich's perturbation theorem [11,13]. In case $\nu(\xi) \sim |\xi|^{\alpha}$ as $|\xi| \to \infty$, $0 \leq \alpha < 1$, the $\alpha^{(j)}(\nu)$ will not be analytic around $\gamma = 0$. Nevertheless, we obtain an asymptotic development

$$\alpha^{(j)}(\nu) \sim \sum_{n=1}^{\infty} \alpha_n^{(j)} |\nu|^n, \qquad (1 \leq j \leq 5)$$

where $\alpha_1^{(j)}$ is imaginary and $\alpha_2^{(j)} < 0$. These constants can be computed by formal perturbation theory. They correspond to the adiabatic sound speed and absorption coefficients for low frequency sound waves [5].

The results (8) and (10) extend known results on finite-state velocity models in one dimension [1,2] to the full three-dimensional linearized Boltzmann equation. These theorems are valid in any number of dimensions. Their proofs and related matters will appear in full detail in [3,4].

We end this paper with several open questions.

1) If one has an external force field $F(x,\xi)$, then the linearized Boltzmann equation (1) becomes (assuming unit mass)

$$(13) \qquad \frac{\partial p}{\partial t} + \xi \cdot \mathrm{grad}_x\, p + F \cdot \mathrm{grad}_\xi\, p = \frac{1}{\epsilon} Qp, \quad \lim_{t \downarrow 0} p = f.$$

The extensions of our limit theorems to this case have not been worked out. Nelson [14; p. 77] has results for an equation with the same form as (13) but where Q has a one-dimensional null-space and is not a Boltzmann collision operator.

2) Grad's decomposition (5) of the operator Q stems from a restriction to a class of intermolecular potentials that are physically unnatural. Pao [16] has shown for a large class of more realistic potentials that Q is self-adjoint and $(Q + \alpha I)^{-1}$ is compact and negative definite for $\alpha > 0$. The limit theorems should hold not only in this case but also for any equation with the same form as (1) provided Q is self-adjoint, negative-semi-definite, with an isolated, finite dimensional eigenvalue at zero. Our methods, based on the existence of the eigenvalues $\alpha^{(\gamma)}(\gamma)$, do not seem to go over.

3) We mentioned the asymptotic nonuniqueness of the Navier-Stokes equations, but question whether these equations have any additional properties which single them out as an asymptotic limit.

4) The statements of our results at least make sense for the nonlinear Boltzmann equation. We feel that a fruitful area of research is the study of nonlinear models. Initial work in this direction has been done by Kurtz [12], who proved the analogue of (8) for the Carleman model. This model, however, has the unsatisfactory feature that its Euler equations are trivial $(A \equiv 0)$. A physically more interesting model has been suggested by Godunov and Sultangazin [6; p. 16].

References

[1] R. Ellis and M. Pinsky, Limit theorems for model Boltzmann equations with several conserved quantities, Indiana Univ. Math. J. 23 (1973), 287-307.

[2] __________, Asymptotic equivalence of the linear Navier-Stokes and heat equations in one dimension, J. Diff. Eqns., 1975.

[3] __________, Projection of the Navier-Stokes upon the Euler equations, J. de Math. Pures et Appl., to appear.

[4] __________, The first and second fluid approximations to the linearized Boltzmann equation, J. de Math. Pures et Appl., to appear.

[5] J. Foch and G. Ford, The dispersion of sound in monoatomic gases, in Studies in Statistical Mechanics, vol. 5, North Holland Press (New York, 1970).

[6] S. Godunov and V. Sultangazin, On discrete models of the kinetic Boltzmann equation, Russian Math Surveys, 26(1971), 1-56.

[7] H. Grad, Asymptotic theory of the Boltzmann equation, in Rarified Gas Dynamics, vol. 1, ed. by J. Laurmann, Academic Press (New York, 1963), 26-60.

[8] __________, Asymptotic equivalence of the Navier-Stokes and non-linear Boltzmann equations, Symposia in Applied Math., vol. 17, Amer. Math. Soc. (Providence, 1965), 154-183.

[9] __________, Principles of the kinetic theory of gases, in Handbuch der Physik, vol. 12, ed. by S. Flügge, Springer-Verlag (Berlin, 1958), 205-294.

[10] __________, Solution of the Boltzmann equation in an unbounded domain, Comm. Pure Appl. Math. 18(1965), 345-354.

[11] T. Kato, Perturbation Theory for Linear Operators, Springer-Verlag (New York, 1966).

[12] T. Kurtz, Convergence of sequences of nonlinear operators with an application to gas kinetics, Trans. A. M. S. 186(1973), 259-272.

[13] J. McLennan, Convergence of the Chapman-Enskog expansion for the linearized Boltzmann equation, Physics of Fluids 8(1965), 1580-1584.

[14] E. Nelson, Dynamical Theories of Brownian Motion, Princeton Univ. Press (Princeton, 1967).

[15] B. Nicolaenko, Dispersion laws for plane wave propagation, in The Boltzmann Equation, ed. by F. A. Grünbaum, Courant Inst. (New York, 1971), 125-173.

[16] Y. Pao, Boltzmann collision operator with inverse-power molecular potentials, Courant Inst. preliminary report.

DUAL MULTIPLICATIVE OPERATOR FUNCTIONALS

Richard Griego[1]

Introduction. This note presents a summary of results that will be published elsewhere together with applications to the theory of random evolutions that are not given here.

The notion of a multiplicative operator functional (MOF) of a Markov process was introduced by Pinsky [8] and [9]. MOF's were motivated by earlier work of Griego and Hersh [4] on random evolutions. Indeed, [4] had pointed out the multiplicative property of random evolutions of Markov chains and introduced the expectation semigroup of a random evolution. MOF's provide realizations and generalizations of random evolutions. MOF's appear to be innocuous generalizations of the now familiar concept of real valued multiplicative functionals of Markov processes as developed, for example, in [1]. However, MOF's (and random evolutions) provide a unified model for many concrete problems that arise in such diverse fields as transport theory, wave propagation in random media, operations research, and systems of partial differential equations.

This note explores dual notions for MOF's. Duality for random evolutions of Markov chains have been studied by Keepler [6] and Schay [10] from a different point of view.

Theorem 3.4 gives the main duality result. However, Theorem 1.2 presents a foundational result on the strong continuity of the expectation semigroup of an MOF that is of interest in itself.

---

(1) Research supported by NSF Grant GP-31811.

1. Multiplicative operator functionals. We use the notation of [1] for Markov processes and related concepts. In what follows let $X = (\Omega,\mathcal{F},\mathcal{F}_t,x(t),\theta_t,P_x)$ be a right continuous Markov process with state space $(E,\mathcal{B})$. We assume E is a separable locally compact metric space with $\mathcal{B}$ the σ-algebra of Borel sets on E. For convenience we assume that all the Markov processes considered in this paper are non-terminating, that is, with infinite lifetimes, although our results are easily modified for finite lifetimes. We also let L be a Banach space with norm $|\cdot|$, and denote by $\mathcal{L}$ the space of bounded linear operators on L and let $\|\cdot\|$ be the operator norm for elements in $\mathcal{L}$. We emphasize that the space L need not be related to the process X in any way. We use the notation $< f^*, f >$ for the value of $f^*(f)$ where $f \in L$, $f^* \in L^*$ and L^* is the adjoint of L.

DEFINITION 1.1. A multiplicative operator functional (MOF) $M = \{M(t),\ t \geq 0\}$ of the pair (X,L) is a mapping $(t,\omega) \to M(t,\omega)$ of $[0,\infty) \times \Omega \to \mathcal{L}$ so that,

(i) there exist finitely valued $\mathcal{F}_t$-measurable operators $M_n(t,\omega) = \sum_{k=1}^{m_n} M_k^{(n)} I_{A_k^{(n)}}(\omega)$, $M_k^{(n)} \in \mathcal{L}$, $A_k^{(n)} \in \mathcal{F}_t$, so that $M_n(t,\omega)f \to M(t,\omega)f$ as $n \to \infty$ a.e. P_x in L, for each $t \geq 0$, $f \in L$, $x \in E$;

(ii) $t \to M(t,\omega)f$ is weakly right continuous a.e. P_x for each $f \in L$ and $x \in E$, that is, $< f^*,M(\cdot,\omega)f >$ is right continuous in $t \geq 0$, a.e. P_x for all $f^* \in L^*$, $f \in L$ and $x \in E$;

(iii) $M(0,\omega)f = f$ a.e. P_x, for each $f \in L$ and $x \in E$;

(iv) $M(t+s,\omega)f = M(t,\omega)M(s,\theta_t\omega)f$ a.e. P_x, for each $x \in E$, $s,t \geq 0$ and $f \in L$.

Note that the order of the operators on the right hand side of condition (iv) is important since the operators need not commute.

An MOF is said to be continuous (CMOF) if the mapping in condition (ii) is weakly continuous in t, a.e. P_x for each $x \in E$.

In the usual definition of an MOF, condition (ii) is strengthened to strong right continuity. However, for our purposes weak continuity suffices and is more convenient. Also, if $\|M(t,\omega)\| \leq K$, a.e. P_x for each $x \in E$, for $t \in (\alpha,\beta)$, $0 < \alpha < \beta < \infty$, where K depends only on α, β and x , then one can adapt Theorem 10.2.3 of [5] to obtain strong continuity. If an MOF happens to satisfy the strong right continuity condition, we then refer to it as a <u>strong MOF</u>.

Let ξ be a σ-finite measure on the state space $(E,\mathcal{B})$. We denote by $L_p(E,L,\xi)$, for $1 \leq p < \infty$, the usual space of equivalence classes of mappings $\varphi: E \to L$ that are strongly measurable (with respect to ξ) and for which $\int_E |\varphi(x)|^p \xi(dx) < \infty$. Each such space is a Banach space relative to the norm $N_p(\varphi) = \left(\int_E |\varphi(x)|^p \xi(dx)\right)^{1/p}$. If L is a reflexive Banach space, $1 < p < \infty$ and $\frac{1}{p} + \frac{1}{q} = 1$, then the adjoint space of $L_p(E,L,\xi)$ is isometrically isomorphic to $L_q(E,L^*,\xi)$ and the correspondence is given as follows: $F \in (L_p(E,L,\xi))^*$ corresponds to $\psi \in L_q(E,L^*,\xi)$ by the formula

$$(1.1) \qquad < F,\varphi > = \int_E < \psi(x),\varphi(x) > \xi(dx) \ , \quad \varphi \in L_p(E,L,\xi) \ .$$

See [3, p. 606] for a discussion of these facts.

Given a Markov process X with state space $(E,\mathcal{B})$ and a measure ξ on $(E,\mathcal{B})$ we write ξP_t for the measure given by

$$(1.2) \qquad \xi P_t(A) = \int_E P(t,x,A)\xi(dx) \ ,$$

where $A \in \mathcal{B}$, $t \geq 0$, and $P(t,x,A) = P_x(x(t) \in A)$.

We say ξ is an <u>excessive measure</u> iff ξ is σ-finite and $\xi P_t \leq \xi$ for all $t \geq 0$.

The following theorem that we present without proof introduces an important parameterization obtained by taking the expectation of an MOF. That the expectation of an MOF determines a semigroup was already noted in [4].

THEOREM 1.1. Let $M = \{M(t),\ t \geq 0\}$ be an MOF of (X,L). Assume ξ is an excessive measure on $(E,\mathcal{B})$ and also

$$(1.3)\qquad (C(t))^p \equiv \sup_{x\in E} (E_x[\|M(t)\|^q])^{p/q} < \infty ,$$

where p is fixed, $1 \leq p < \infty$ and $\frac{1}{p} + \frac{1}{q} = 1$. For $\varphi \in L_p(E,L,\xi)$ define

$$(1.4)\qquad T(t)\varphi(x) = E_x[M(t)\varphi(x(t))] .$$

Then $\{T(t),\ t \geq 0\}$ is a semigroup of bounded linear operators on $L_p(E,L,\xi)$ and $\|T(t)\| \leq C(t)$.

The semigroup given by (1.3) is called the expectation semigroup of M.

The following result gives conditions for the strong continuity of the expectation semigroup. The proof will be given elsewhere.

THEOREM 1.2. Let $M = \{M(t),\ t \geq 0\}$ be an MOF of (X,L) and ξ an excessive measure on $(E,\mathcal{B})$. Assume M satisfies (1.3). Assume further that L is a reflexive Banach space, and that for a fixed $\delta > 0$,
$C \equiv \sup_{0\leq t\leq \delta} \sup_{x\in E} (E_x[\|M(t)\|^q])^{p/q} < \infty$, where $\frac{1}{p} + \frac{1}{q} = 1$, and p is fixed, $1 < p < \infty$. Then (1.4) defines a strongly continuous semigroup $\{T(t),\ t \geq 0\}$ of bounded linear operators on $L_p(E,L,\xi)$.

2. Dual Markov processes. Given a right continuous Markov process $X = (\Omega,\mathcal{F},\mathcal{F}_t,x(t),\theta_t,P_x)$ with state space $(E,\mathcal{B})$, the resolvent $\{U^\alpha,\ \alpha \geq 0\}$ of X is given by

$$(2.1)\qquad U^\alpha f(x) = E_x \int_0^\infty e^{-\alpha t} f(x(t))dt = \int_0^\infty e^{-\alpha t} P_t f(x)dt ,$$

for $x \in E$, $f: E \to R$ bounded and $\mathcal{B}$-measurable, and where $P_t f(x) = E_x[f(x(t))]$ gives the semigroup of X. Let $U^\alpha(x,dy)$ be the measure $U^\alpha(x,A) = U^\alpha I_A(x)$ for $A \in \mathcal{B}$, so that $U^\alpha f(x) = \int_E f(y)U^\alpha(x,dy)$.

In what follows we will consider two standard non-terminating Markov processes $X = (\Omega,\mathcal{F},\mathcal{F}_t,x(t),\theta_t,P_x)$ and $\hat{X} = (\hat{\Omega},\hat{\mathcal{F}},\hat{\mathcal{F}}_t,\hat{x}(t),\hat{\theta}_t,\hat{P}_x)$ on the same state space $(E,\mathcal{B})$. See [1] for the definition of a standard process. We take Ω and $\hat{\Omega}$ to both be the canonical space of maps $\omega: [0,\infty) \to E$ that are right continuous with left limits, and $x(t,\omega) = \hat{x}(t,\omega) = \omega(t)$ and $\theta_t\omega = \theta_t\hat{\omega}$.

We say that X and $\hat{X}$ are in duality with respect to the σ-finite measure ξ on $(E,\mathcal{B})$ if for each $\alpha > 0$ the following conditions are satisfied:

(a) the measures $U^\alpha(x,\cdot)$ and $\hat{U}^\alpha(x,\cdot)$ are absolutely continuous with respect to ξ for each $x \in E$; and

(b)

$$\int_E f(x)\cdot U^\alpha g(x)\xi(dx) = \int_E \hat{U}^\alpha f(x)\cdot g(x)\xi(dx) . \tag{2.2}$$

These conditions imply the existence of a density $u^\alpha(x,y)$ so that

$$\begin{aligned} U^\alpha f(x) &= \int u^\alpha(x,y)f(y)\xi(dy) , \quad \text{and} \\ \hat{U}^\alpha f(y) &= \int u^\alpha(x,y)f(x)\xi(dx) . \end{aligned} \tag{2.3}$$

The left and right hand sides of (2.2) are the Laplace transforms of $\int f(x)P_t g(x)\xi(dx)$ and $\int g(x)\hat{P}_t f(x)\xi(dx)$, respectively, and since these integrals are right continuous in t for continuous f and g, they are equal. We write the equality of these two integrals in the form

$$E_\xi[f(x(0))g(x(t))] = \hat{E}_\xi[g(\hat{x}(0))f(\hat{x}(t))] \tag{2.4}$$

where $E_\xi[\quad] = \int E_x[\quad]\xi(dx)$ and a similar expression for $\hat{E}_\xi$.

It is known that the measure ξ is excessive, [1, Cor. 1.12, p. 259]. The dual process $\hat{X}$ is thought of roughly as the process X run backwards in time, but it is difficult to make this notion precise. There are various papers on certain reversed Markov processes with the direction of time explicitly reversed. These results are not in a form that are easily applicable to the issues at hand in this paper and the relationship between dual processes and reversed processes is not complete. See [2] and [7] for results and applications of reversed processes. We prefer to use the above setup of dual processes in order to study duality for MOF's. We will carry out this study by means of a device introduced by Walsh [11] that allows one, so to speak, to continuously reverse the time parameter in a process and in this manner construct functionals for the dual process $\hat{X}$ from functionals of X.

Following Walsh then, we introduce the reversal operator r_t as follows: let X and $\hat{X}$ be as above and fix $t > 0$; we say two elements ω and ω' of Ω are t-<u>equivalent</u> if $\omega(s) = \omega'(s)$ for all s, $0 \leq s \leq t$. The reversed operator r_t is defined on t-equivalence classes $\bar{\omega}$ in a manner so that if $\omega \in \bar{\omega}$, then $r_t\bar{\omega}$ is the set of elements ω' in Ω satisfying

$$(2.5) \qquad \omega'(s) = \omega(t-s-0) \quad \text{if} \quad s < t .$$

We also define $r_0\omega = \omega$ for all ω. By abuse of notation, if we write $r_t\omega$ for ω' then we can write (2.5) as $r_t\omega(s) = \omega(t-s-0)$ or $x(s,r_t\omega) = x(t-s-0,\omega)$ for $s < t$. Note that each $\omega'(\cdot)$ is itself a right continuous path with left limits; this is due to taking the limit from the left at s in (2.5).

We will find the following fact about r_t useful:

$$(2.6) \qquad \begin{aligned} &(a) \quad r_{s+t}\omega \text{ is t-equivalent to } r_t\theta_s\omega , \text{ and} \\ &(b) \quad r_s\omega \text{ is s-equivalent to } \theta_t r_{s+t}\omega . \end{aligned}$$

Walsh's paper [11, p. 236] has an instructive diagram that indicates the truth of these statements.

Suppose that $\alpha(t,\omega)$ is a real valued function of $t \geq 0$ and $\omega \in \Omega$ so that $\alpha(t,\cdot)$ is $\mathcal{F}_t$- measurable for each $t \geq 0$. We say that $\alpha(t,\omega)$

is path adapted iff whenever ω and ω' are t-equivalent then $\alpha(s,\omega) = \alpha(s,\omega')$ for all $s < t$. If $\alpha(t,\omega)$ is path adapted then for a fixed t, $\alpha(t) \circ r_t$ is well defined since $\alpha(t,\omega)$ is constant on t-equivalence classes. This extends as well to the case where $\alpha(t,\omega)$ is Banach space valued. Path adaptedness is satisfied by a functional $\alpha(t,\omega)$ if $\alpha(t,\cdot)$ is $\mathcal{F}^0_{t-}$-measurable, where $\mathcal{F}^0_{t-} = \sigma(\bigcup_{\epsilon>0} \mathcal{F}^0_{t-\epsilon})$ and $\mathcal{F}^0_t = \sigma(x(s), 0 \leq s \leq t)$. If $t \to \alpha(t,\omega)$ is a.s. left continuous then the $\mathcal{F}^0_t$ - measurability of $\alpha(t,\cdot)$ suffices. Since the σ-algebras $\mathcal{F}_t$ are obtained from the $\mathcal{F}^0_t$ by a process of completion, the property of path adaptedness need not be satisfied by a general (even continuous) $\mathcal{F}_t$ - measurable functional $\alpha(t,\omega)$. The MOF's encountered in the applications below can always be taken to be path adapted.

In general, given a function Y on Ω, $Y \circ r_t$ is defined to be any function Z on Ω so that if $\omega \in \Omega$ and if $\bar{\omega}$ is the t-equivalence class containing ω, there then exists $\omega' \in r_t\bar{\omega}$ for which $Z(\omega) = Y(\omega')$. In general, this definition determines $Y \circ r_t$ only up to sets of $\hat{P}_\xi$- measure zero. In fact, it is shown in [11, Thm. 2.1] that if Y is measurable with respect to the P_ξ - completion of $\mathcal{F}^0_t$ for some $t > 0$, then $Y \circ r_t$ is measurable with respect to the $\hat{P}_\xi$ - completion of $\mathcal{F}^0_t$ and, furthermore,

$$E_\xi[Y] = \hat{E}_\xi[Y \circ r_t] \; . \tag{2.7}$$

We note by the definition of a standard process, that if Y is measurable with respect to $\mathcal{F}_t$, then Y satisfies the above requirement of being measurable with respect to the P_ξ - completion of $\mathcal{F}^0_t$. We will have occasion to apply (2.7) only to path adapted Banach space valued functionals.

3. Dual multiplicative operator functionals. In what follows we fix two standard processes X and $\hat{X}$ with state space $(E,\mathcal{B})$ and in duality with respect to ξ. We also fix a Banach space L and denote its dual space by L^*. Strong continuity and measurability properties regarding L^* are to be understood with respect to the topology on L^* induced by the

norm $|f^*| = \sup\{|< f^*,f >| : |f| \leq 1 , f \in L\}$. A function $t \to \psi^*(t)$ of $[0,\infty)$ to L^* is said to be w^*-continuous iff $\lim_{s\to t} < \psi^*(s),f > = < \psi^*(t),f >$ for each $t \geq 0$, $f \in L$, that is, if $\psi^*(t)$ is a continuous function of t with respect to the w^*- topology on L^* . Similarly, a mapping $\eta^*: \Omega \to L^*$, where $(\Omega,\mathcal{F})$ is a measurable space, is defined to be w^*- $\mathcal{F}$ - measurable if $\omega \to < \eta^*(\omega),f >$ is a real valued $\mathcal{F}$ - measurable random variable for each $f \in L$.

It is shown in [11, Thm. 4.1 and Prop. 4.5] that if $m(t,\omega)$ is a real-valued (path adapted) continuous multiplicative functional of X , then $\hat{m}(t) = m(t) \circ r_t$ defines a continuous multiplicative functional of $\hat{X}$. We wish to adapt this result to our case of operator valued functionals. Thus, let $M = \{M(t), t \geq 0\}$ be a path adapted CMOF of (X,L). We define $\hat{M} = \{\hat{M}(t), t \geq 0\}$ as follows:

$$\hat{M}(t) = (M(t) \circ r_t)^* , \tag{3.1}$$

where * denotes the operator dual of the random linear operator in the parentheses.

THEOREM 3.1. $\hat{M}$ satisfies the following properties:

(a) $\hat{M}(t,\omega)$ is a bounded linear operator on L^* for each $t \geq 0$ and $\omega \in \Omega$;

(b) $\omega \to \hat{M}(t,\omega)f^*$ is w^* - $\hat{\mathcal{F}}_t$- measurable for each $f^* \in L^*$ and $t \geq 0$;

(c) $t \to \hat{M}(t,\omega)f^*$ is w^*- continuous a.e. P_x , for each $f^* \in L^*$, $t \geq 0$, and $x \in E$;

(d) $\hat{M}(0,\omega)f^* = f^*$ a.e. P_x , for each $f^* \in L^*$ and $x \in E$;

(e) $\hat{M}(t+s,\omega)f^* = \hat{M}(t,\omega)\hat{M}(s,\theta_t\omega)f^*$ a.e. P_x , for each $s,t \geq 0$, $f^* \in L^*$, and $x \in E$.

We give only the proof of the multiplicative property (e) which is as follows: we have a.s. that,

$$\begin{aligned}\hat{M}(s+t) &= (M(t+s) \circ r_{t+s})^{*} \\ &= (M(t) \circ r_{t+s}\, M(s) \circ \theta_t r_{t+s})^{*} \\ &= (M(s) \circ \theta_t r_{t+s})^{*}(M(t) \circ r_{t+s}) \\ &= (M(s) \circ r_s)^{*}\, (M(t) \circ r_t \theta_s)^{*} \qquad \text{(by (2.6))} \\ &= \hat{M}(s)\hat{M}(t) \circ \theta_s .\end{aligned}$$

DEFINITION 3.2. A family of operators $\hat{M} = \{\hat{M}(t),\ t \geq 0\}$ satisfying the properties (a) - (e) of Theorem 3.1 is called a w^{*}- continuous multiplicative operator functional (w^{*}- CMOF) of $(\hat{X},L^{*})$. We will call the w^{*}- CMOF of $(\hat{X},L^{*})$ defined by (3.1) the w^{*}- dual of the CMOF, $M = \{M(T),\ t \geq 0\}$ of (X,L).

The following corollary is an easy consequence of Theorem 3.1.

COROLLARY 3.3. If L is a separable, reflexive Banach space then the w^{*} - dual $\hat{M}$ of M is a CMOF of $(\hat{X},L^{*})$.

The next theorem states that under appropriate conditions the expectation semigroups of M and $\hat{M}$ are dual semigroups.

THEOREM 3.4. Let L be a separable, reflexive Banach space and let $\hat{M} = \{\hat{M}(t),\ t \geq 0\}$ be the dual of $M = \{M(t),\ t \geq 0\}$ as above. Assume for some $\delta_1, \delta_2 > 0$ that $\sup_{0 \leq t \leq \delta_1} \sup_{x \in E} E_x^{p/q}[\|M(t)\|^q] < \infty$ and $\sup_{0 \leq t \leq \delta_2} \sup_{x \in E} \hat{E}_x^{q/p}[\|\hat{M}(t)\|^p] < \infty$, where $1 < p < \infty$, $\frac{1}{p} + \frac{1}{q} = 1$. Then the expectation semigroup $\{\hat{T}(t),\ t \geq 0\}$ of $\hat{M}$ given by $\hat{T}(t)\psi(x) = \hat{E}_x[\hat{M}(t)\psi(\hat{x}(t))]$ for $\psi \in L_q(E,L^{*},\xi)$, is well defined and strongly continuous on $L_q(E,L^{*},\xi)$, and furthermore, $< \hat{T}(t)\psi,\varphi > = < \psi,T(t)\varphi >$ for all $\varphi \in L_p(E,L,\xi)$ and $\psi \in L_q(E,L^{*},\xi)$, that is, $\hat{T}(t) = T^{*}(t)$, the operator dual of $T(t)$.

Proof. By Theorem 1.2 and Corollary 3.3 $\{\hat{T}(t),\ t \geq 0\}$ is strongly continuous on $L_q(E,L^*,\xi)$. Also,

$$\begin{aligned}
< \hat{T}(t)\psi,\varphi > &= \int_E < \hat{T}(t)\psi(x),\varphi(x) > \xi(dx) \\
&= \int_E < \hat{E}[\hat{M}(t)\psi(\hat{x}(t))],\ \varphi(x) > \xi(dx) \\
&= \int_E \hat{E}[< \hat{M}(t)\psi(\hat{x}(t)),\varphi(x) >]\ \xi(dx) \\
&= \hat{E}_\xi[< \hat{M}(t)\psi(\hat{x}(t+)),\varphi(\hat{x}(0)) >] \\
&\qquad (\hat{x}(t+) = \hat{x}(t)\text{: right continuity of paths}) \\
&= \hat{E}_\xi[< \psi(\hat{x}(t+)),M(t) \circ r_t\varphi(\hat{x}(0)) >] \\
&= \hat{E}_\xi[< \psi(\hat{x}(0)),M(t)\varphi(\hat{x}(t)) > \circ\ r_t]^{(1)} \\
&= E_\xi[< \psi(x(0)),M(t)\varphi(x(t)) >] \quad (\text{by } (2.7)) \\
&= \int_E E_x[< \psi(x(0)),M(t)\varphi(x(t)) >]\ \xi(dx) \\
&= \int < \psi(x),E_x[M(t)\varphi(x(t))] > \xi(dx) \\
&= < \psi,T(t)\varphi > .
\end{aligned}$$

Q.E.D.

BIBLIOGRAPHY

1. R.M. Blumenthal and R.K. Getoor, Markov processes and potential theory, Academic Press, New York, 1968.

2. K.L. Chung and J.B. Walsh, To reverse a Markov process, Acta. Math., 123 (1969), 225-251.

3. R.E. Edwards, Functional analysis; Holt, Rinehart and Winston, New York, 1965.

4. R. Griego and R. Hersh, Theory of random evolutions with applications to partial differential equations, Trans. Amer. Math. Soc., 156 (1971), 405-418.

5. E. Hille and R.S. Phillips, Functional analysis and semi-groups, rev. ed., Amer. Math. Soc. Colloq. Publ., vol. 31, Amer. Math. Soc., Providence, R.I., 1957.

6. M. Keepler, Backward and forward equations for random evolutions, Doctoral dissertation, University of New Mexico, 1973.

7. H. Kunita and T. Watanabe, On certain reversed processes and their applications to potential theory and boundary theory, J. Math. Mech., 15, No. 3 (1966), 393-434.

8. M.A. Pinsky, Multiplicative operator functionals of a Markov process, Bull. Amer. Math. Soc., 77 (1971), 377-380.

9. M.A. Pinsky, Multiplicative operator functionals and their asymptotic properties, Advances in probability, vol. 3, 1-100, Marcel Dekker, New York, 1974.

10. G. Schay, Forward equations for random evolutions, preprint, Univ. of Massachusetts, 1973.

11. J.B. Walsh, Markov processes and their functionals in duality, Z. Wahrs. verw. Geb., 24 (1972), 229-246.

Vol. 277: Séminaire Banach. Edité par C. Houzel. VII, 229 pages. 1972. DM 22,-

Vol. 278: H. Jacquet, Automorphic Forms on GL(2) Part II. XIII, 142 pages. 1972. DM 18,-

Vol. 279: R. Bott, S. Gitler and I. M. James, Lectures on Algebraic and Differential Topology. V, 174 pages. 1972. DM 20,-

Vol. 280: Conference on the Theory of Ordinary and Partial Differential Equations. Edited by W. N. Everitt and B. D. Sleeman. XV, 367 pages. 1972. DM 29,-

Vol. 281: Coherence in Categories. Edited by S. Mac Lane. VII, 235 pages. 1972. DM 22,-

Vol. 282: W. Klingenberg und P. Flaschel, Riemannsche Hilbertmannigfaltigkeiten. Periodische Geodätische. VII, 211 Seiten. 1972. DM 22,-

Vol. 283: L. Illusie, Complexe Cotangent et Déformations II. VII, 304 pages. 1972. DM 27,-

Vol. 284: P. A. Meyer, Martingales and Stochastic Integrals I. VI, 89 pages. 1972. DM 18,-

Vol. 285: P. de la Harpe, Classical Banach-Lie Algebras and Banach-Lie Groups of Operators in Hilbert Space. III, 160 pages. 1972. DM 18,-

Vol. 286: S. Murakami, On Automorphisms of Siegel Domains. V, 95 pages. 1972. DM 18,-

Vol. 287: Hyperfunctions and Pseudo-Differential Equations. Edited by H. Komatsu. VII, 529 pages. 1973. DM 40,-

Vol. 288: Groupes de Monodromie en Géométrie Algébrique. (SGA 7 I). Dirigé par A. Grothendieck. IX, 523 pages. 1972. DM 55,-

Vol. 289: B. Fuglede, Finely Harmonic Functions. III, 188. 1972. DM 20,-

Vol. 290: D. B. Zagier, Equivariant Pontrjagin Classes and Applications to Orbit Spaces. IX, 130 pages. 1972. DM 18,-

Vol. 291: P. Orlik, Seifert Manifolds. VIII, 155 pages. 1972. DM 18,-

Vol. 292: W. D. Wallis, A. P. Street and J. S. Wallis, Combinatorics: Room Squares, Sum-Free Sets, Hadamard Matrices. V, 508 pages. 1972. DM 55,-

Vol. 293: R. A. DeVore, The Approximation of Continuous Functions by Positive Linear Operators. VIII, 289 pages. 1972. DM 27,-

Vol. 294: Stability of Stochastic Dynamical Systems. Edited by R. F. Curtain. IX, 332 pages. 1972. DM 29,-

Vol. 295: C. Dellacherie, Ensembles Analytiques Capacités. Mesures de Hausdorff. XII, 123 pages. 1972. DM 18,-

Vol. 296: Probability and Information Theory II. Edited by M. Behara, K. Krickeberg and J. Wolfowitz. V, 223 pages. 1973. DM 22,-

Vol. 297: J. Garnett, Analytic Capacity and Measure. IV, 138 pages. 1972. DM 18,-

Vol. 298: Proceedings of the Second Conference on Compact Transformation Groups. Part 1. XIII, 453 pages. 1972. DM 35,-

Vol. 299: Proceedings of the Second Conference on Compact Transformation Groups. Part 2. XIV, 327 pages. 1972. DM 29,-

Vol. 300: P. Eymard, Moyennes Invariantes et Représentations Unitaires. II, 113 pages. 1972. DM 18,-

Vol. 301: F. Pittnauer, Vorlesungen über asymptotische Reihen. VI, 186 Seiten. 1972. DM 18,-

Vol. 302: M. Demazure, Lectures on p-Divisible Groups. V, 98 pages. 1972. DM 18,-

Vol. 303: Graph Theory and Applications. Edited by Y. Alavi, D. R. Lick and A. T. White. IX, 329 pages. 1972. DM 26,-

Vol. 304: A. K. Bousfield and D. M. Kan, Homotopy Limits, Completions and Localizations. V, 348 pages. 1972. DM 29,-

Vol. 305: Théorie des Topos et Cohomologie Etale des Schémas. Tome 3. (SGA 4). Dirigé par M. Artin, A. Grothendieck et J. L. Verdier. VI, 640 pages. 1973. DM 55,-

Vol. 306: H. Luckhardt, Extensional Gödel Functional Interpretation. VI, 161 pages. 1973. DM 20,-

Vol. 307: J. L. Bretagnolle, S. D. Chatterji et P.-A. Meyer, Ecole d'été de Probabilités: Processus Stochastiques. VI, 198 pages. 1973. DM 22,-

Vol. 308: D. Knutson, λ-Rings and the Representation Theory of the Symmetric Group. IV, 203 pages. 1973. DM 22,-

Vol. 309: D. H. Sattinger, Topics in Stability and Bifurcation Theory. VI, 190 pages. 1973. DM 20,-

Vol. 310: B. Iversen, Generic Local Structure of the Morphisms in Commutative Algebra. IV, 108 pages. 1973. DM 18,-

Vol. 311: Conference on Commutative Algebra. Edited by J. W. Brewer and E. A. Rutter. VII, 251 pages. 1973. DM 24,-

Vol. 312: Symposium on Ordinary Differential Equations. Edited by W. A. Harris, Jr. and Y. Sibuya. VIII, 204 pages. 1973. DM 22,-

Vol. 313: K. Jörgens and J. Weidmann, Spectral Properties of Hamiltonian Operators. III, 140 pages. 1973. DM 18,-

Vol. 314: M. Deuring, Lectures on the Theory of Algebraic Functions of One Variable. VI, 151 pages. 1973. DM 18,-

Vol. 315: K. Bichteler, Integration Theory (with Special Attention to Vector Measures). VI, 357 pages. 1973. DM 29,-

Vol. 316: Symposium on Non-Well-Posed Problems and Logarithmic Convexity. Edited by R. J. Knops. V, 176 pages. 1973. DM 20,-

Vol. 317: Séminaire Bourbaki – vol. 1971/72. Exposés 400–417. IV, 361 pages. 1973. DM 29,-

Vol. 318: Recent Advances in Topological Dynamics. Edited by A. Beck. VIII, 285 pages. 1973. DM 27,-

Vol. 319: Conference on Group Theory. Edited by R. W. Gatterdam and K. W. Weston. V, 188 pages. 1973. DM 20,-

Vol. 320: Modular Functions of One Variable I. Edited by W. Kuyk. V, 195 pages. 1973. DM 20,-

Vol. 321: Séminaire de Probabilités VII. Edité par P. A. Meyer. VI, 322 pages. 1973. DM 29,-

Vol. 322: Nonlinear Problems in the Physical Sciences and Biology. Edited by I. Stakgold, D. D. Joseph and D. H. Sattinger. VIII, 357 pages. 1973. DM 29,-

Vol. 323: J. L. Lions, Perturbations Singulières dans les Problèmes aux Limites et en Contrôle Optimal. XII, 645 pages. 1973. DM 46,-

Vol. 324: K. Kreith, Oscillation Theory. VI, 109 pages. 1973. DM 18,-

Vol. 325: C.-C. Chou, La Transformation de Fourier Complexe et L'Equation de Convolution. IX, 137 pages. 1973. DM 18,-

Vol. 326: A. Robert, Elliptic Curves. VIII, 264 pages. 1973. DM 24,-

Vol. 327: E. Matlis, One-Dimensional Cohen-Macaulay Rings. XII, 157 pages. 1973. DM 20,-

Vol. 328: J. R. Büchi and D. Siefkes, The Monadic Second Order Theory of All Countable Ordinals. VI, 217 pages. 1973. DM 22,-

Vol. 329: W. Trebels, Multipliers for (C, α)-Bounded Fourier Expansions in Banach Spaces and Approximation Theory. VII, 103 pages. 1973. DM 18,-

Vol. 330: Proceedings of the Second Japan-USSR Symposium on Probability Theory. Edited by G. Maruyama and Yu. V. Prokhorov. VI, 550 pages. 1973. DM 40,-

Vol. 331: Summer School on Topological Vector Spaces. Edited by L. Waelbroeck. VI, 226 pages. 1973. DM 22,-

Vol. 332: Séminaire Pierre Lelong (Analyse) Année 1971-1972. V, 131 pages. 1973. DM 18,-

Vol. 333: Numerische, insbesondere approximationstheoretische Behandlung von Funktionalgleichungen. Herausgegeben von R. Ansorge und W. Törnig. VI, 296 Seiten. 1973. DM 27,-

Vol. 334: F. Schweiger, The Metrical Theory of Jacobi-Perron Algorithm. V, 111 pages. 1973. DM 18,-

Vol. 335: H. Huck, R. Roitzsch, U. Simon, W. Vortisch, R. Walden, B. Wegner und W. Wendland, Beweismethoden der Differentialgeometrie im Großen. IX, 159 Seiten. 1973. DM 20,-

Vol. 336: L'Analyse Harmonique dans le Domaine Complexe. Edité par E. J. Akutowicz. VIII, 169 pages. 1973. DM 20,-

Vol. 337: Cambridge Summer School in Mathematical Logic. Edited by A. R. D. Mathias and H. Rogers. IX, 660 pages. 1973. DM 46,-

Vol: 338: J. Lindenstrauss and L. Tzafriri, Classical Banach Spaces. IX, 243 pages. 1973. DM 24,-

Vol. 339: G. Kempf, F. Knudsen, D. Mumford and B. Saint-Donat, Toroidal Embeddings I. VIII, 209 pages. 1973. DM 22,-

Vol. 340: Groupes de Monodromie en Géométrie Algébrique. (SGA 7 II). Par P. Deligne et N. Katz. X, 438 pages. 1973. DM 44,-

Vol. 341: Algebraic K-Theory I, Higher K-Theories. Edited by H. Bass. XV, 335 pages. 1973. DM 29,-

Vol. 342: Algebraic K-Theory II, "Classical" Algebraic K-Theory, and Connections with Arithmetic. Edited by H. Bass. XV, 527 pages. 1973. DM 40,-

Vol. 343: Algebraic K-Theory III, Hermitian K-Theory and Geometric Applications. Edited by H. Bass. XV, 572 pages. 1973. DM 40,-

Vol. 344: A. S. Troelstra (Editor), Metamathematical Investigation of Intuitionistic Arithmetic and Analysis. XVII, 485 pages. 1973. DM 38,-

Vol. 345: Proceedings of a Conference on Operator Theory. Edited by P. A. Fillmore. VI, 228 pages. 1973. DM 22,-

Vol. 346: Fučík et al., Spectral Analysis of Nonlinear Operators. II, 287 pages. 1973. DM 26,-

Vol. 347: J. M. Boardman and R. M. Vogt, Homotopy Invariant Algebraic Structures on Topological Spaces. X, 257 pages. 1973. DM 24,-

Vol. 348: A. M. Mathai and R. K. Saxena, Generalized Hypergeometric Functions with Applications in Statistics and Physical Sciences. VII, 314 pages. 1973. DM 26,-

Vol. 349: Modular Functions of One Variable II. Edited by W. Kuyk and P. Deligne. V, 598 pages. 1973. DM 38,-

Vol. 350: Modular Functions of One Variable III. Edited by W. Kuyk and J.-P. Serre. V, 350 pages. 1973. DM 26,-

Vol. 351: H. Tachikawa, Quasi-Frobenius Rings and Generalizations. XI, 172 pages. 1973. DM 20,-

Vol. 352: J. D. Fay, Theta Functions on Riemann Surfaces. V, 137 pages. 1973. DM 18,-

Vol. 353: Proceedings of the Conference on Orders, Group Rings and Related Topics. Organized by J. S. Hsia, M. L. Madan and T. G. Ralley. X, 224 pages. 1973. DM 22,-

Vol. 354: K. J. Devlin, Aspects of Constructibility. XII, 240 pages. 1973. DM 24,-

Vol. 355: M. Sion, A Theory of Semigroup Valued Measures. V, 140 pages. 1973. DM 18,-

Vol. 356: W. L. J. van der Kallen, Infinitesimally Central-Extensions of Chevalley Groups. VII, 147 pages. 1973. DM 18,-

Vol. 357: W. Borho, P. Gabriel und R. Rentschler, Primideale in Einhüllenden auflösbarer Lie-Algebren. V, 182 Seiten. 1973. DM 20,-

Vol. 358: F. L. Williams, Tensor Products of Principal Series Representations. VI, 132 pages. 1973. DM 18,-

Vol. 359: U. Stammbach, Homology in Group Theory. VIII, 183 pages. 1973. DM 20,-

Vol. 360: W. J. Padgett and R. L. Taylor, Laws of Large Numbers for Normed Linear Spaces and Certain Fréchet Spaces. VI, 111 pages. 1973. DM 18,-

Vol. 361: J. W. Schutz, Foundations of Special Relativity: Kinematic Axioms for Minkowski Space Time. XX, 314 pages. 1973. DM 26,-

Vol. 362: Proceedings of the Conference on Numerical Solution of Ordinary Differential Equations. Edited by D. Bettis. VIII, 490 pages. 1974. DM 34,-

Vol. 363: Conference on the Numerical Solution of Differential Equations. Edited by G. A. Watson. IX, 221 pages. 1974. DM 20,-

Vol. 364: Proceedings on Infinite Dimensional Holomorphy. Edited by T. L. Hayden and T. J. Suffridge. VII, 212 pages. 1974. DM 20,-

Vol. 365: R. P. Gilbert, Constructive Methods for Elliptic Equations. VII, 397 pages. 1974. DM 26,-

Vol. 366: R. Steinberg, Conjugacy Classes in Algebraic Groups (Notes by V. V. Deodhar). VI, 159 pages. 1974. DM 18,-

Vol. 367: K. Langmann und W. Lütkebohmert, Cousinverteilungen und Fortsetzungssätze. VI, 151 Seiten. 1974. DM 16,-

Vol. 368: R. J. Milgram, Unstable Homotopy from the Stable Point of View. V, 109 pages. 1974. DM 16,-

Vol. 369: Victoria Symposium on Nonstandard Analysis. Edited by A. Hurd and P. Loeb. XVIII, 339 pages. 1974. DM 26,-

Vol. 370: B. Mazur and W. Messing, Universal Extensions and One Dimensional Crystalline Cohomology. VII, 134 pages. 1974. DM 16,-

Vol. 371: V. Poenaru, Analyse Différentielle. V, 228 pages. 1974. DM 20,-

Vol. 372: Proceedings of the Second International Conference on the Theory of Groups 1973. Edited by M. F. Newman. VII, 740 pages. 1974. DM 48,-

Vol. 373: A. E. R. Woodcock and T. Poston, A Geometrical Study of the Elementary Catastrophes. V, 257 pages. 1974. DM 22,-

Vol. 374: S. Yamamuro, Differential Calculus in Topological Linear Spaces. IV, 179 pages. 1974. DM 18,-

Vol. 375: Topology Conference 1973. Edited by R. F. Dickman Jr. and P. Fletcher. X, 283 pages. 1974. DM 24,-

Vol. 376: D. B. Osteyee and I. J. Good, Information, Weight of Evidence, the Singularity between Probability Measures and Signal Detection. XI, 156 pages. 1974. DM 16,-

Vol. 377: A. M. Fink, Almost Periodic Differential Equations. VIII, 336 pages. 1974. DM 26,-

Vol. 378: TOPO 72 – General Topology and its Applications. Proceedings 1972. Edited by R. Alò, R. W. Heath and J. Nagata. XIV, 651 pages. 1974. DM 50,-

Vol. 379: A. Badrikian et S. Chevet, Mesures Cylindriques, Espaces de Wiener et Fonctions Aléatoires Gaussiennes. X, 383 pages. 1974. DM 32,-

Vol. 380: M. Petrich, Rings and Semigroups. VIII, 182 pages. 1974. DM 18,-

Vol. 381: Séminaire de Probabilités VIII. Edité par P. A. Meyer. IX, 354 pages. 1974. DM 32,-

Vol. 382: J. H. van Lint, Combinatorial Theory Seminar Eindhoven University of Technology. VI, 131 pages. 1974. DM 18,-

Vol. 383: Séminaire Bourbaki – vol. 1972/73. Exposés 418-435 IV, 334 pages. 1974. DM 30,-

Vol. 384: Functional Analysis and Applications, Proceedings 1972. Edited by L. Nachbin. V, 270 pages. 1974. DM 22,-

Vol. 385: J. Douglas Jr. and T. Dupont, Collocation Methods for Parabolic Equations in a Single Space Variable (Based on C^1-Piecewise-Polynomial Spaces). V, 147 pages. 1974. DM 16,-

Vol. 386: J. Tits, Buildings of Spherical Type and Finite BN-Pairs. IX, 299 pages. 1974. DM 24,-

Vol. 387: C. P. Bruter, Eléments de la Théorie des Matroïdes. V, 138 pages. 1974. DM 18,-

Vol. 388: R. L. Lipsman, Group Representations. X, 166 pages. 1974. DM 20,-

Vol. 389: M.-A. Knus et M. Ojanguren, Théorie de la Descente et Algèbres d' Azumaya. IV, 163 pages. 1974. DM 20,-

Vol. 390: P. A. Meyer, P. Priouret et F. Spitzer, Ecole d'Eté de Probabilités de Saint-Flour III – 1973. Edité par A. Badrikian et P.-L. Hennequin. VIII, 189 pages. 1974. DM 20,-

Vol. 391: J. Gray, Formal Category Theory: Adjointness for 2-Categories. XII, 282 pages. 1974. DM 24,-

Vol. 392: Géométrie Différentielle, Colloque, Santiago de Compostela, Espagne 1972. Edité par E. Vidal. VI, 225 pages. 1974. DM 20,-

Vol. 393: G. Wassermann, Stability of Unfoldings. IX, 164 pages. 1974. DM 20,-

Vol. 394: W. M. Patterson 3rd, Iterative Methods for the Solution of a Linear Operator Equation in Hilbert Space – A Survey. III, 183 pages. 1974. DM 20,-

Vol. 395: Numerische Behandlung nichtlinearer Integrodifferential- und Differentialgleichungen. Tagung 1973. Herausgegeben von R. Ansorge und W. Törnig. VII, 313 Seiten. 1974. DM 28,-

Vol. 396: K. H. Hofmann, M. Mislove and A. Stralka, The Pontryagin Duality of Compact O-Dimensional Semilattices and its Applications. XVI, 122 pages. 1974. DM 18,-

Vol. 397: T. Yamada, The Schur Subgroup of the Brauer Group. V, 159 pages. 1974. DM 18,-

Vol. 398: Théories de l'Information, Actes des Rencontres de Marseille-Luminy, 1973. Edité par J. Kampé de Fériet et C. Picard. XII, 201 pages. 1974. DM 23,-